Oil Mortality in Post-Fossil Fuel Era Nigeria

Augustine Sadiq Okoh

Oil Mortality in Post-Fossil Fuel Era Nigeria

Beyond the Oil Age

Springer

Augustine Sadiq Okoh
Department of Political Science
Benue State University
Makurdi, Nigeria

ISBN 978-3-030-60784-5 ISBN 978-3-030-60785-2 (eBook)
https://doi.org/10.1007/978-3-030-60785-2

This Springer imprint is published by the registered company Springer Nature Switzerland AG
The registered company address is: Gewerbestrasse 11, 6330 Cham, Switzerland

This book is dedicated to a gentleman and officer, Major General Johnston Olubummi Irefin, the General Officer Commanding the 6th Division of Nigerian Army for his commitments to the world of military and academic excellence.

Preface

Nigeria is prone to cyclical distortions in the international market. Uncertainties of the oil market make the resource considered as the "heartbeat of the nation" also to be the country's Achilles heel. Nigeria's response to the perennial volatility in the oil market is largely one of ambivalence and not commensurate with scale of the problem. In most cases, economic diversification has become a rhetorical statement conjured in moments of crisis to assuage the restive society. But once the crisis is temporally overcome, the planners of national development seem to relapse to the status quo. These recurrent lapses show that the efforts are mainly grandstanding commitment to economic diversification, and hence was never meant to redirect the sinking ship of Nigerian economy.

There is no tangible clean or green energy strategy and growth plans comparable to those articulated in other oil-producing countries like Saudi Arabia. Saudi Arabia which is the world's foremost exporter of crude oil in her Vision 2030 plan laid out concrete steps to reduce her dependence on a single commodity. The country's economic diversification strategy strives to wean her fossil fuel dependence. They correctly assumed oil faces an uncertain future, hence the need for oil production and renewable energy proliferation to exist side by side. To this end, the development plan laid out concrete steps of gradually phasing out oil from the energy supply mix with this initiative as the main driver of the unbundling of Arabian American Oil Company (ARAMCO).

A systemic green growth strategy similar to those fashioned in Ethiopia and Rwanda targeted at optimizing vast opportunities in renewable energy and electric mobility to transform their energy supply sector has not been initiated in Nigeria. Equally, strategies similar to Kenya's energy plan of improving total renewable energy generation from 70% in 2018 to 100% of the total power needs by 2020 are nonexistent in our government's blueprint. Rather, the SE4ALL plan targets optimization of renewable in 2030 with limited inroad made toward energy security. Though there are hotchpotch policies with different sectoral policies and institutional frameworks focused on promoting low carbon growth. Yet, they do not prioritize activities for scale-up investments in renewables comparative to those planned for Kenyan and Saudi Arabian societies. These countries have placed strong faith in alternative energies to drive their green growth agenda and in so doing will generate additional revenue from other sources when a crash in oil prices occurs. To meet this goal, appropriate regulatory frameworks coupled with an effective implementation platform have been designed to harness the vast renewable energy potentials thereby curbing fossil fuels from their national energy supply mix.

It is not totally a sorry state in Nigeria's economic diversification agenda. Nigerian government is not totally insensitive to rising need for weaning her hydrocarbon addiction. Different governments have initiated differing biofuel production but has so far not yielded any tangible benefit. In recognition of the looming crisis, the Vice-President, Prof. Yemi Osinbajo in an address to the 55th Anniversary of Oil Producers Trade Section of the Lagos Chambers of Commerce and Industry (LCCI) held on November 2, 2017 made a passionate appeal for economic diversification. Professor Osinbajo in his opening speech envisioned oil looming danger resulting from its elimination from energy supply mix. He berated Nigeria's reliance on a single commodity which to him forebodes grave danger if adequate measures are not taken to broaden the country's revenue base. The Vice-President then charged oil stakeholders to make adequate arrangements for a future without oil given that exit from fossil fuel "is no longer a question of if but when."

In fulfillment of the Vice-president's apocalyptic vision, crude oil prices crashed in April 2020 to less than US $10 per barrel. A combination of human and non-human factors led to the dramatic crash, prominent among which is the Coronavirus (COVID-19) pandemic of 2019 and the Russia/Saudi Arabia price impasse. Concurrently, a crash of the global capital market occurred with international stock valuations for the NSE-ASI, Nikkei, Dow Jones, and FTSE-100 declining by an average of 23.8% between January and March 2020 while Global Airlines lost about US $252 billion in revenues. Consequently, different sectors of the global economy from hospitality to services recorded massive losses due to COVID-19 incited lockdowns. In the wake of the slump, the International Monetary Fund (IMF) predicts a gloomy future ahead for most nations seeing that the global economy would decline by 3% in 2020.

Again, Nigeria has not been left off the hook. Unemployment figure in April 2020 resulting from COVID-19 according to the National Bureau of Statistic (NBS) was 33.6%. According to the NBS, economic growth could fall by as much as −4.40% to −8.91% depending on the length of the lockdown period. In recognition of the economic hardship further lockdown entails, government sketched a recovery strategy appropriately titled: Bouncing Back: Nigeria Economic Sustainability Plan (NESP) aimed at rekindling the dying flames of the economy as well as ameliorating adverse effects of COVID-19. NESP proposes to improve foreign exchange, reduce unemployment, and drive sustainable growth. To achieve this, the document seeks to increase capital flow, generate new jobs, and improve agricultural production while also promoting local content to stifle capital flight as well as catalyze innovations. The purpose is to kill two birds with one stone: growing the economy to weather the vagaries of COVID-19 while also using low carbon development to meet her Nationally Determined Contribution (NDC) to the Paris Agreement. The ultimate goal is to turn the downward spiral of economic growth using low carbon development as its springboard.

Though government formulated a post-COVID-19 recovery plan that will build an inclusive economy, some inversely related issues are still undermining these objectives. These inversely relational issues sit at the cross-road of three national cross-cutting crises: climate change, economic/health insecurity, and energy shortfall. All these encumbrances are pulling the economy toward divergent ends. Nigeria's inversely related and competing needs are growing in opposite directions. But this should not be the case. It is an antithetical contradiction to be growing the economy using fossil as echoed in the NESP documents while surreptitiously reducing Greenhouse Gas (GHG) intensity of the economy. Specifically, this book realigns Nigeria's growing need for decarbonization and economic recovery by plotting a green economic agenda to a decarbonized polity mindful of the looming end of oil in global economies. The aim is to point toward other viable options Nigeria can explore in traveling to a post-petroleum civilization. In so doing, the book simultaneously fashions a carbon-neutral pathway out of the present economic/public health debacle. This prompts the book to ask and answer the following questions:

What are the main drivers of curbing fossil fuel from the global energy supply mix and how prepared is Nigeria to run emergent economic and human health existential threats race of improving human security while reducing energy intensity of the economy in the post-COVID-19 age?

Is Nigeria's carbon-led development on an ecologically neutral path in the post-fossil world?

Are there structural, political, and economic tools propelling Nigeria toward meeting the Nationally Determined Contribution (NDC) and Sustainable Development Goals (SDGs) such that there is trade-off between these mutually exclusive goals in making Nigeria's post-fossil fuel society green?

Is the economic stimulus package sufficient in meeting rising challenges in economic terrain while also ensuring clean energy transition thereby reordering the country's outsize carbon budget in a finite world where natural resources distribution weighs down by human decisions?

What are the viable options for building Nigeria's post-carbon society and curbing threats of climate change/COVID-19 while maintaining Nigeria's economic viability such that the nation lives within bounds of the natural world?

In the following ten chapters, we shall vigorously unravel an eco-efficient roadmap needed surmount the gathering storm when the petroleum bubble burst. Though government formulated policies to meet her Paris Agreement commitments, these measures cannot put a foot in the door of her reduction of the carbon-based power generation neither can it provide additional income for the country when the petroleum windfall winds down. Diversifying Nigeria's energy asset base such that the non-resource sector grows to generate jobs once the oil reserves winds up or the oil windfall winds down should be a critical priority of government in the post-petroleum era. Such a policy should align with the sustainable development goals, Nigeria Economic Sustainability Plan (NESP), NDC, and long-term post-Vision 20:2020 targets of the country.

The book owes an intellectual debt to Prof. Kayoade Shoremeku of the Covenant University, Ota. In an inaugural lecture, Prof. Shoremeku berated Nigeria for its inconsistent handling of the vast oil wealth which to him fulfills the apocalyptic prophesy of Zaki Yamani. Yamani, who was once a Minister of Petroleum in Saudi Arabia at the peak of oil euphoria in 1973 prophetically cautioned that the stone age did not end because the world ran out of stone, neither will the oil age end because of lack of oil. Given oil's transience, Professor Shoremeku suggests a book on this subject is waiting to be written and will probably be titled: Oil and the Mortality of the Nigerian State. This book takes Shoremeku's challenge headlong. It speaks to Nigeria's oil mortality and uses green economics to expose the gaps before proffering solutions for embarking upon these uncertainties. The book's primary goal is to rebalance production, reprioritize consumption, and enshrine allocative justice. So, its primary focus is to promote sustainable resource utilization by exploring other alternatives for building a climate-resilient economy while ameliorating public health challenges in a post-COVID-19 world.

Ultimately, then, this book serves as a toolkit for plotting an environmentally benign path out of present (and future) economic, health, and environmental crises. The book is therefore intended as a valuable resource for students, Civil Society Organizations, policy-makers, and academics who are interested in finding a carbon-neutral path out of Nigeria's economic *cul-de-sac* fostered by the decarbonization of the global energy economy. The findings of this study should trigger a national debate on the looming exit from fossil fuels. In doing so, it will—hopefully—reduce the impact of coronavirus, accelerate integration of low carbon development into the Nigerian national development plan while opening up the country to new investment for building a carbon-neutral society mindful of potential pitfalls from health-related impediments. Lessons learnt from Nigeria's precarious circumstance will be of immense benefit to other oil prospecting, oil-producing, and non-producing nations who seek an equitable way of pursuing two inversely related goals in the post-COVID-19 era.

Makurdi, Nigeria Augustine Sadiq Okoh

Acknowledgements

Debts of varied dimensions accrue to several persons in writing this book. There are many friends and colleagues who in one way or another took time from their busy schedules to read and comment on the book. Their comments and encouragements were of great help.

My first gratitude goes to Prof. Kayoade Shoremeku for stimulating my academic interest in the transience nature of oil and need for finding new viable alternatives to oil. Next is Prof. Chuks Okereke of University of Reading, UK whom I started this book project with. But due to other pressing international commitment as lead author, IPCC Sixth Assessment Report could not contribute to the book. His insightful comments and criticisms at the formative stage of this book immensely added value to the book. Professor John Igoli of Federal University of Agriculture, Makurdi also commented and contributed in editing this work and must be commended for his selfless service. Not to forget my friend Dr. Sam Ogallah of Solidaridad International for his contributions of editing the book.

Finally, my thanks go to Jessica Fleming, County Councillor (Cons) Hartismere, Suffolk, for her interest in my academic works. Our years of LinkedIn contact nurtured into an academic friendship which culminated in her being an important part of this work. Within a short period of time, Jessica professionally edited the book—*pro bono*—bringing it to the standard it is today. Jess, words cannot express the depth of my gratitude.

Observations

- Nigeria's post-fossil fuel civilization sits at the crossroads of three interrelated challenges: climate change, economic, and technological encumbrances.
- There is the paradox of stabilizing greenhouse gas emission at 1.5 °C of the Paris Climate Agreement while also ensuring a strong economy in post-COVID-19 era.
- Minimal synergy exists between different national economic blueprints and the NDC.
- Hotchpotch policies with different sectoral policies and institutional frameworks are focused on promoting low carbon growth without prioritizing activities in the form of scale-up of investments in renewables.
- Nigeria's energy system is shaped by decisions on decarbonization, production systems, and investment priorities which are intermeshed with political manoeuvrings in an increasingly artificial, crowded, and changing world where the planet is coming together and drifting apart.
- Though activities on economic diversification are on the rise, political inertia resulting from oil stakeholders' hold on power railroaded the energy transition toward maintaining the status quo.
- Technological shift in energy transition has not been fully explored to improve energy access neither is their interrelatedness explored.
- Traditional energy systems are in silos with no linkages with the new sets of energy actors, and hence sector coupling to foster energy security is minimal.
- Nigeria's economic diversification is on a free-spin with no form of standardization, and hence economic slumps will slow the pace of development as nominal overtures are made toward renewable energy alternative.
- There is no wide range of national capacity for the MRV system to conduct transparent and verifiable GHG emissions across sectors in Nigeria.
- Streamlining of the new NDC with the NESP is fraught with lapses and gaps.
- To meet the NDC obligation requires embracing new norms and values as well as new forms of technologies and climate support services capable of meeting emerging needs in a changing world.

Key Takeaways for Government

- Create institutional linkages between NDC and other national and international priorities such as the Sustainable Development Goals and Agricultural Transformation Agenda, NESP to engender climate resiliency.
- Promote political lock-in on economic diversification through awareness program aimed at reorientation of society toward phasing out fossil fuels while also ameliorating COVID-19 pandemic.
- Domestic laws such as the Petroleum Industry Governance Bill (PIGB) adopted and enforced to stem further uncertainties in the downstream sector should be fully implemented.
- Designate a responsible agency or institution comprising a monitoring and evaluation unit with the authority to coordinate all issues of the NDC and formulate an NDC implementation index and monitoring/tracking tools.
- Promote climate-smart agriculture education at all levels where the importance of bamboos and other biomass as alternative energy source to fossil fuels is not only desirable but a way to go.
- Synergistic integration of bamboo cultivation with other bio-energy policy is carried out to address food and energy crop paradox in the biofuel policy.
- Create disincentive in the form of eco-tax and eco-label for fossil fuel industry and incentive for private sector-led investments through granting subsidy to the clean energy sector.
- Formulate new legislation on green economy growth termed the National Low Carbon and Growth Plan (NLCGP) for the country.
- Create a "fee-and-dividend" system to impose carbon fee on fossil fuel companies with 100% of the revenue collected distributed monthly to the people on a per capita basis as dividend.
- Sustain the Green Bond initiative of the Federal Government and promote youth, women, and people with disability involvement in climate actions in Nigeria.
- Create a Just Transition Taskforce (JTT) to keep track of progress while also ensuring alignment of NDC with NESP.

Contents

List of Figures

List of Tables

List of Maps

Chapter 1
Dawn of a New Age

> We will not go back to normal. Normal never was. Our pre-corona existence was not normal other than we normalized greed, inequality, exhaustion, depletion, extraction, disconnection, confusion, rage, hoarding, hate and lack. We should not long to return, my friends. We are being given the opportunity to stitch a new garment. One that fits all of humanity and nature.
> Brene Brown
> "[T]he Stone Age did not end for lack of stone, and the Oil Age will end long before the world runs out of oil."
> Sheikh Zaki Yamani (Formerly Saudi Arabian Oil Minister) *the Economist*, 23 October 2003

The Gathering Storm

Dramatic events are causing structural change in the global energy economy. In the last five years transformation within the energy sector is causing a technological shift to occur. Within the global energy economy a new wave of proliferation of low carbon technologies are receiving serious attention as of late. These transformations are mainly in energy access and energy transition. The two concepts are complexly intertwined with consumption and production of crude oil. Their rising profile in the international political economy owe to their crucial role of providing energy needed for economic activities hence are tied to the economic fortunes of the nation state. So, the major driver of oil production in oil producing state is the urge to use their abundant petroleum resources to meet developmental requirements. This is because an efficient utilization of petroleum resources entails energy efficient technology which is equally dependent on revenue accruing from crude oil to propagate clean energy technologies. While the former needs a stable environment for crude oil production to make any meaningful change, the latter is the upshot of two closely interrelated global instruments: Sustainable Development Goals (SDGs) and the Paris climate agreement.

A. S. Okoh, *Oil Mortality in Post-Fossil Fuel Era Nigeria*,
https://doi.org/10.1007/978-3-030-60785-2_1

Critical component of energy access and energy transition is not just the large scale energy production; it is more of the fact that one promotes economic growth while the other insists on decarbonization using efficient technologies. Though these issues have divergent goals, they have one thing in common: they are highly susceptible to external pressure. Certainly, in the oil industry international price fluctuation is material to how the economy makes use of income derived from the trade in crude oil.

Despite this paradox, transformation of the global energy economy is receiving serious attention lately. This transformation is part of unequivocal commitment of the international community to address the trilemma of climate change, food security and energy poverty. **For sure,** Clean Energy Transition (CET) is part of the decarbonization measures initiated to avert dangerous climate change. It strives to address the climate emergency stringently targeting keeping the global average temperature below 1.5 °C threshold considered safe for planetary existence. Limiting the global mean temperature to below 2 °C is in line with Article 2 of the UN Framework Convention on Climate Change (UNFCCC). Article 2 calls for a reduction of GHG concentration in the atmosphere to 'a level that would prevent dangerous anthropogenic interference with the climate system'. Conversely, the Sustainable Development Goals seek to end poverty in all its forms (water, food and energy poverty) most especially goal #7 targeted at ensuring universal access to affordable, reliable, sustainable, and modern energy for all.

The world community's objective here is to enshrine a culture of efficiency in energy consumption and production. Needless to say, transformation of the global economy will have different ramifications for different societies. What we can emphatically say here is how this transformation is not only in the energy sector since it is also happening within the public health services. The driver of public health services transformation as the World Health Organization (WHO) fittingly observed is the novel coronavirus pandemic of 2019 (COVID-19). COVID-19 pandemic is not only a public health emergency; it is human tragedy in that it is the foremost threat to human existent in recent memory. Coronavirus pandemic has not only created misery, it brought the world to one of its darkest hours given that it has caused loss of many lives. The number of coronavirus linked causality is growing exponentially daily. For instance, in June 2020 WHO reported that there were 6.19 m confirmed cases of COVID-19 with 376,000 deaths recorded worldwide. But in July 2020 there was an upsurge in cases of COVID-19 which resulted in over 16 million confirmed cases and about 650,000 deaths worldwide.

Although the transmission of the pandemic from person to person in Africa was slow at the onset, it has picked up lately and this is due mainly to activities of international travelers. As many international travelers are now returning to their countries the cases of the pandemic is rising. Hence the continent which hitherto recorded 157,241 confirmed cases in June 2020 with 4,472 deaths skyrocketed to a total of 780,000 confirmed cases in July 2020 with 16,715 deaths registered across the continent. South Africa tops the number of confirmed cases with 408,052, followed by Nigeria with 39, 977 and 856 deaths in July 2020. The low figure reported in most African countries is not unexpected since these countries have negligible number of testing facilities. Nigeria with a population of over 200 million has a low affliction and mortality rate owing to paucity of data on the actual of number of confirmed cases and the number of people who died from coronavirus related ailment.

COVID-19 impact is not just in causing massive loss of life; it has also exposed the Nigerian economy to vicissitudes of the marketplace. Three direct threats of weighty dimension afflict her in the aftermath of the exposure to COVID-19 pandemic namely: the economic meltdown, food shortages and extreme poverty. The breakdown of public health services also added to severity of these risks. Taken together, these threats have resulted in the loss of life and property, dwindling economic fortunes in the process is reinforcing pauperization of the poor. This is because the oil shock resulting from temporary ban placed on economic activities within and between states slowed down the free movement of goods and services. The result of this temporary ban is that demand side shortfall occasioned by COVID-19 brought oil prices to an all-time low (Fig. 1.1) making crude oil price to reach a negative threshold in April, 2020.

Nigeria's crude oil export dependent economy was not spared from the financial shock linked to the shrinkage in demand. In line with the drop in demand prices for most crude oil grades showed a discount as low as $1.51 and $3.95 for Qua Iboe and Bonny light respectively to future dated Brent in May, 2020. As a result of this the country lost an average of $35 m a day in accruable revenue from April 2020 to June 2020. It is trite to say that the glut triggered by the coronavirus hit the oil industry in Nigeria so hard. That much is obvious since the country recourse to storing on ships and barges which, again, increased the cost of shipping. So, it is right to say that

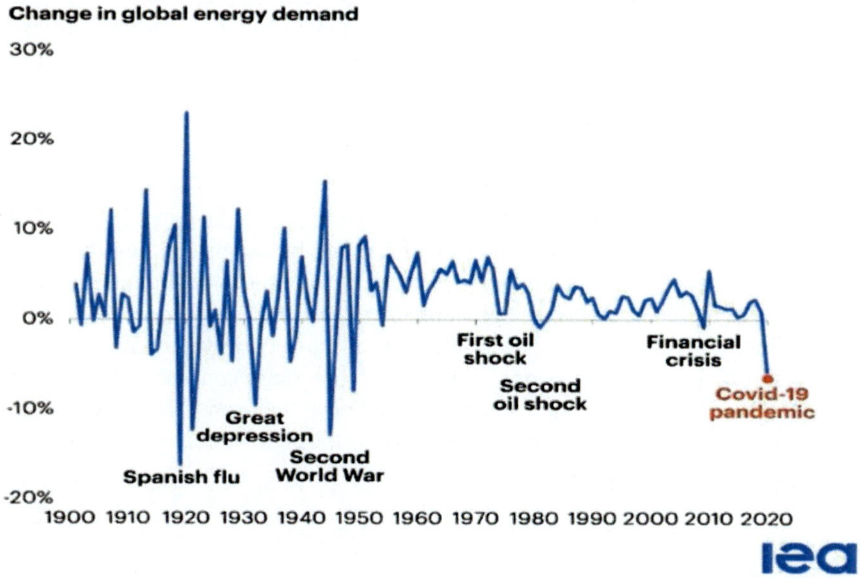

Fig. 1.1 Global energy demand from 1900 to 2020. *Source* IEA (2020)

COVID-19 pandemic brought Nigeria to the lowest ebb of life in that the financial crisis placed severe economic stress on the economy.

With a weakened Naira, the country seems on the road to Venezuela-like collapse of the economy. This collapse is not just because of the rising debt portfolio, but more of the fact that coronavirus-related shocks have all the trappings of a sovereign debt crisis in the offing. The situation is further aggravated by the loans taken to ameliorate economic failure brought about by the pandemic. With dwindling revenue from crude oil, government had to borrow funds from international lenders to meet her 2020 budget commitments. Towards this end, the International Monetary Fund (IMF) granted Nigeria a $3.4 billion emergency loan in April 2020 to cushion the impact of the economic meltdown.

In another twist to the unfolding uncertainty in the global energy market, most mature economies are now hatching strategies phasing out fossil fuel. This ubiquitous shift away from fossil fuels is a direct threat to Nigeria's economic fortunes hence will make the post-COVID-19 economic recovery far more difficult since the country is highly susceptible to shock and stress from the volatile energy market. Moreover, the oil and gas sector is the main income meant for realising near-term developmental needs while also strengthening long term viability of the economy. It is therefore not surprising that despite her pursuit of short-term goal for economic development other measures in the form of Low Carbon Development (LCD) minimally exist side by side. Such coexistence of diverse goals creates the impression of growing in opposite directions. Better still, it may be a case of paying lips service to one of the goals. One would expect that with the trend towards phasing out fossil fuel gaining momentum the country should robustly pursue only sustainable energy objective instead of taking inversely related routes to a decarbonized polity.

It is against the backdrop of the foregoing that one sees Nigeria's carbon politics as a carbon catch-22. Certainly, Nigeria's carbon dilemma is a classic catch-22 condition of pursuing two divergent goals: low carbon development on the heels of economic growth. The implication is the country will require its vast hydrocarbons deposits to build a post-carbon world. Herein lies the rub: building a carbon neutral future which is unmindful of economic diversification will derail the national climate commitment. This is a tough ask requiring policy levers to bail the country from potential economic doldrums and dead-end. If the country should not to be caught holding the wrong end of the environmental sustainability stick then planning for a future without oil must be one of the essential tasks of government in post-COVID-19 era. Government is determined to reduce the emission intensity of the economy by minimizing the primary source of GHG emissions while also promoting economic development. How Nigeria will rise above the conflictive situation and narrow prism of upholding only carbon-led outlook is an onerous task. This is the analytic fulcrum of this study and subject of the subsequent section.

Analytic Hub of the Study

In the past crude oil export was not the main income of pre-independence Nigeria. This fact is obvious to all. What is not too obvious is how crude oil supplanted other commodities in the economy to become the main provider of foreign exchange? What are the issues exacerbating deterioration of the country's economic base? What are the drivers of the new wave of technological shift and what academic tools have been devoted to eliminating uncertainties of the shift from an anthropocentric vision towards a more bio-centric and people-centric relationship with nature?

For starter, agriculture was the mainstay of colonial era economy of Nigeria. In the colonial era policy formulated by the colonialist favoured expropriation of agricultural resources to the citadels of capital in the Northern Hemisphere. But the discovery of oil in 1956 at Oloibiri changed agriculture's prominence as the main foreign exchange earner. In its place crude oil became the main foreign exchange earner. The mercurial rise of oil to national prominence is associated with its intrinsic nature of generating immense wealth. On this ground, extensive attention was paid to oil at the detriment of other sectors with the hope it can engender endogenous growth. They also erroneously assumed that oil can always bail the economy from its malfeasance including its internal contradictions. In some climes oil is the engine of growth fuelling massive infrastructural development to meet the needs of a growing economy. That, unfortunately, is not the case here. In the face of rising uncertainties attached to the commodity there is an acceptance within academic and political circles that oil cannot take Nigeria to the promised economic nirvana since it is not totally void of problems. Combinations of intravenous and extraneous factors have conspired to erode the commodity's significance but its major undoing is its own internal contradictions which has weighed it down.

One of the major issues is it is pollution intensive. Another issue is it susceptibility to price fluctuations at the international market. It is a known fact—at least in Nigeria—that limited level of economic certainty exists when oil revenue is benchmark of economic development. Oil's vital role in the economy together with the need for meeting the global decarbonization obligation enmeshed the country in a paradoxical condition. It is therefore not surprising that economic diversification which is supposed to reduce reliance on a single commodity has generated debates within political and academic circles. Vigorous debates over exploitation of nature, the use of vast natural resources to build an industrial economy and exploring new alternatives to reduce reliance on a single product are now growing academic pastime.

On one side, there are those clamoring for the country to grow the economy through abundant resources while on other side are others who hold that oil should play a critical role but there must be an avenues for exploring trade-off between economic and environmental sustainability. However, the debate is not limited to economic sustainability and geophysical factors alone but has biophysical undercurrent including emergent challenge in the form of post-COVID-19 economic recovery strategy. It is therefore not unexpected that series of controversies trail these cross-cutting issues since they all have different champions pull government towards ideological ends. This divergence instead of pointing towards a constructive and objective path towards economic rationalization and diversification rather mystifies than points to an alternative source of income. A positive attribute is there are different solutions proffered from these debates even then no concrete inroad has been made towards building a post-carbon civilization. The dilemma facing the country now is not only how to surmount the deficiencies cropping up from socioeconomic challenges within the states, but how to rise above the current rebranding of the internalisation of economization of nature to a more bio-centric view. Though the main accelerants of these crises have international coloration, internal contradictions resulting weak institutions created avenues for the replication of the vicious cycle of pillage of resources. With doubts attached to government's revenue, development projects are the first sacrificial victims of this policy blind sight. The worrying thing here is the nation is in need of inclusive and eco-efficient solutions to reverse perennial crisis of global economic depression. Still, no headway has been made to frontally tackle this. Such a solution must be based on an empirical study of the crisis to establish its potential for meeting economic diversification needs thus stimulating a positive change.

From the foregoing, the real concern today is not just addressing the COVID-19 pandemic but surmounting the tug-of-war between economic growth and green growth. Judging from the facts on the ground, Nigeria cannot come to grips with the uncertainty arising from navigating the thin line between economic growth and de-growth (materialization and de-materialization) while also protecting long-term decarbonisation targets. As a matter of fact, there are very many grey areas in the plan of action stipulated in the Nationally Determine Contribution (NDC) and Nigeria Economic Sustainability Plan (NESP) since politics rather than ecological consideration is the major determinant of the country's policies.

Given doubts arising from sustainability of oil production the country is in need of a new compass to plot a course through the muddy waters of unending price slumps while also building a low carbon economy. Reversing the perennial crisis of global economic depression to one where the economy is capable of withstanding an uneven terrain is the goal of this work. Such endeavor is a herculean task. We do not claim to have all the answers but surmise that some of the issues will unfurl within the tight confines of this book. Our solution is based on an empirical study of the crisis in order to establish the actual depth of the crisis. In this vane, we examine the forces undermining development and the potentials of the proposed solutions stimulating a positive change in a way Nigeria can benefit from opportunities presented in the global decarbonization obligation. For now, limited studies delved into this perplexing issue using green economics to empirically rationalize the issues. To be sure, none of these studies frontally tackle the crisis with a view of fleshing out an eco-sufficient green economic blueprint capable of leading to an ecological nirvana. Instead, oil is still the dominant driver rationalized to drive growth with no other viable option on the table. Rather there are allusions to biofuel but with no concrete treatise on a common and easily accessible crop.

It is against the backdrop of the imminent collapse of oil industry and want of visible alternative that some have argued that Nigerians are living on a borrowed time, with borrowed robes waiting for inevitable economic and political meltdown. Though this may sound farfetched there is merit in it since oil is a known source of conflict. Where conflict exists, there is allocative crisis or unequal exchange which is a governance failure hence will erode gains of sustainable development. Thus conflict over oil revenue is a recipe for crisis. The book goes beyond this economic rationalization of conflict to situate Nigeria's socio-environmental crisis in its proper political milieu where conflicts between different forces struggling to subvert nature are prevalent. As one might expect, the analysis of conflict is not limited to the flow and depletion of natural resources alone. Indeed, the analysis of conflict transcends this simplistic view of man/nature dualisms but delves into the real issues behind resource conflicts alongside the unending circles of oil shocks. Here, questions are asked: who uses these resources? When? For what purpose is it used? At what cost? And, with what impact?

It is trite to say that the 'black gold' transformed Nigeria in a fundamental way is an incontrovertible fact. Equally factual is the assertion that it is also the Achilles Heel of the country. Based on this shifting role where one cannot pin any definite form to oil one would agree it is not a sustainable solution for future generations. Yes, oil is the way forward for present generation but cannot be the way out of economic insecurity for future generation in that it is highly susceptible to boom-bust cycle with a serious dose of inconsistency impairing realization of its immense benefits.

Our goal is to draw an analytic portrait of Nigeria's future without oil, establishing the basic stylized facts surrounding COVID-19 import for national development and transition to a post-fossil fuel society before pointing an equitable footpath towards a carbon neutral world. It is generally acceptable that tackling the twin challenge of climate change and public health encumbrances will jumpstart economic growth. Interest in low carbon development is growing with the likes of Christiana Figures championing the call for exploring eco-efficient alternatives at the same with implementing green economic recovery strategies in post-COVID-19 era. On this ground Nigeria Economic sustainability Plan (NESP) of 2020 was formulated with a view to promoting economic recovery concurrently making concessions toward Low Carbon Development (LCD).

In the following nine chapters, we shall critically examine these issues considering the quest to exit from fossil fuels, economic diversification, the global commitment to reduce GHG emissions and need for a green economic recovery for the COVID-19 era. Contrary to expectations of these interrelated issues pointing towards a constructive path to an equitable ecological distribution and improvement in human and environmental well-being, it mystifies rather than point towards an alternative path for ending our fossil fuel obsession. One will expect a new blueprint for addressing the crisis would be formulated by now. But on the contrary, power strategies are being conceived within different regions for total control of oil. Instead of this struggle over the source of wealth opening a new revolutionary pathway for a sustainable future for the country, it has become an avenue for promoting personal or ethnic stake.

To bring into focus these complex issues the book adopted the green economy perspectives. Our choice hinges on green economics' nature of being a powerful analytic tool for unearthing both economic growth and environmental needs while also ensuring intergeneration equity with future generations. The concept's attribute of being both a policy outlook as well as an academic discipline made it endearing as a suitable toolkit to bring conceptual clarity to an issue with an overwhelming complex nature. Green economy is a concept with a holistic lens of looking at the economy. Within the fold of green economy, the green economic variant is our preferred variant due to its attribute of ensuring carbon production and consumption act within biospheric limit of a country. Adoption of this variant points us towards other viable alternatives effectual in diversifying not just the energy supply base but also shows other options for eradicating Nigeria's oil dependence. Although a large pool of academic literature exists on economic diversification, limited number of these studies is aligned with Nigeria's post-COVID era needs, decarbonisation commitments and economic diversification goals. Inability to find a common ground for resolving these issues created apathy about the way out of Nigeria's economic diversification debacle.

In joining ongoing economic diversification debate, this study examines themes and frames within the ambit of green economics. Our goal is to point not only towards an alternative rationalization of nature which can be operationalized without drawbacks. It is one that leverages abundant resources to equitably meet Nigeria's developmental requirements within the framework of the global decarbonization mandate. This is not say the pace recorded in human development and impressive GDP of 7% in 2019 will be eroded in transiting to a low carbon economy. Far from it! What this objectifies is, the present roadmap is somehow ineffective but that is not said that it has no merit. To a large extent what this shows is the need for finding an approach which strives for a balance between consumption and production of resources.

It has been argued how the present economic blueprint does not serve the people neither does it serve nature's wellbeing. Hence it is viewed to have failed the country since it did not point towards viable options capable of making the post-fossil fuel society energy efficient. The green economics concept adopted here is not only a reference term for scholars interested in transitioning to an ecologically sustainable polity. It goes beyond such mundane rationalization to a more far reaching ramification as both a concept and toolkit for governing low carbon growth. In the book, we extensively examined these intricacies together with the role extreme weather events play in re-conceptualisation of the relationship between human and non-human nature. In particular, the current strategy targeted at minimizing hydrocarbons' role in the economy has both biogeophysical and economic imperative. No doubt, climate extreme events have altered how we relate with one another. No less important is the fact that its interference is not limited to our economic activities alone given that it is also interfering with our conceptualisation of the toolkits needed to bring ecological sanity to the global economic uncertainties.

More importantly, the contribution of this book to knowledge stems from using green economics principles to find ways out of Nigeria's looming environmental and health crises. Beyond this, the book's major contribution is plotting a new pathway out of Nigeria's economic gridlock. Ultimately, then, the book fashions new avenues of exiting fossil fuel which are not injurious to Nigeria's economic wellbeing but ensures a stable environment for human health as well. It goes beyond current emphasis of green economics to find an African-owned rationalization of nature. It use Africa's bio-communitarianism context as the principal focus since this aspect of green economics aligns us with traditional African cultural norms and values.

It is pertinent to draw attention to the fact that dealing with Nigeria's undue dependence on oil is an urgent task decision-makers must undertake. If Nigeria should be successful in eliminating fears associated with climate change together with the need for economic development in the new age, then, it is imperative we adopt an approach which speaks to African norms and values. Such rationalization must be within the framework of internationally acceptable economic and ecological values. This toolkit is a combination of policy/regulatory framework that guides consumption, extraction and technology deployed to limit our pillage of nature. To be sure, this toolkit will limit gas flaring and oil spills while also ensuring stable growth within the energy sector. More critically, it will reduce climate-related risk and its consequential impact of decelerating economic development. The failure to find an eco-sufficient approach is the handicap of the current handling of our economic diversification. The adopted analytic tool will show us potential pitfalls in our reliance on fossil fuel together with shortcomings of the GHG reduction plan encapsulated in the NDC whilst ensuring internal contradictions do not stifle economic development nor impair Nigeria's international obligations.

This is against the backdrop of the prevalence of an acute dose of fiscal discipline imbalance to transform revenue into assets capable of generating income to be consumed and not relying on a single resource. The assumption in most economies is to find other viable alternatives where the nation has comparative advantage. Sorry to say, finding other options is not on the front burners of government's plan in that oil play's a mercurial role of generating revenue. The erroneous assumption is not in isolation. It is an outcrop of the assumption in academic and political circles of how fossil fuel depletion will reach a threshold where Nigeria will be forced to find other alternatives. But the reverse is the case now. Contrary to expectation of fossil fuel assets diminishing it is growing from strength to strength with new oil wells being discovered. Nigeria's interest in finding new oil wells and not opening up new alternatives to fossil fuel is the country's albatross.

This is why: Nigerian economy is highly susceptible to external shocks given that the present low-income base severely undermines the country's adaptive capacity. Accordingly, any external shock will have severe economic repercussion. For instance, the external shock foisted by the price war between Russia and Saudi Arabia in 2020 together with the COVID-19 pandemic imposed further strains on the fragile economy with the vulnerability varying from region to region. Alas, the income base of Nigeria changed with the severity of the pandemic thus exposing the country's susceptibility and inadequacy in confronting the scourge. In the following section an appraisal of the emerging world order is made. This is with a view to pointing us towards the interdependent crises Nigeria's petroleum-led economy shall be grappling with in post-carbon and COVID-19 world.

Cross-Cutting Challenge of Modern Age

A trilemma of interrelated issues bedevils Nigeria's oil dependent economy pushing her further away from meeting national and international obligations. On one hand, there is the complexity of tackling climate change without sacrificing economic growth. Then, there is the challenge of sustainable utilization of natural resources capable of transforming the economy while also ensuring remarkable improvement in environmental quality. Another critical issue is the contradictions of meeting Nigeria's international commitment encapsulated in the Nationally Determined Contribution (NDC) stipulating gradual weaning from fossil fuels while simultaneously using revenue from crude oil to improve the living standards of the people. All these issues occur in multiple chains at diverse scales pulling the economy towards divergent ends. So we ask: How did these complexities come into being? What are the drivers of this multi-scale crisis and why it is important to forge a common ground for these issues?

In answering these questions it is important we established nexus of the crises Nigeria is grappling with. For starters, the 2015 is a landmark year in global climate governance. It is milestone because the Paris Agreement and SDGs were all fleshed out by the world community. Thereafter, the world witnessed rapid structural changes in energy production and consumption. Correspondently, the pace of deployment of new technologies increased to tally with the urgency attached to finding solutions for the twin challenge of climate change and food-energy poverty. The pace of proliferation was too fast that it became known in some quarters as the 'innovation tsunami' or Third Industrial Revolution. The Third Industrial Revolution or the 'innovation Tsunami' did not happen by chance neither did it drop from a parachute. It is the outcrop of conscious policy to tackle in concrete terms existential crisis that beleaguer humanity.

A major factor propelling ongoing technological shift is climate change. Another consequential accelerant of the crisis is volatility of the oil market pushing the global economy towards circles of perennial uncertainties. A case in point is the bombing of Saudi Arabia's oil processing plant in September 2019 which caused over 20% rise in cost of oil in just one day. Another instance is the recent Russia and Saudi Arabian oil price wars of 2020. The Coronavirus crisis of 2020 also added a new twist to the crisis seeing that the pandemic also caused oil price to reach a negative threshold.

In recognition of its severity to economic growth, the world community has taken a common stand to eradicate most avoidable sources of pollution as well as minimize if not totally eliminate exhaustion of earth's finite resources. This is a tall order. Nevertheless, global actions to bring than the level of Greenhouse Gases (GHG) reached an appreciable level. This effort culminated in the re-engineering of energy production in the process is radically altering the landscape of proliferations of new technologies.

Prior to this, Nigeria had high expectations that her vast oil resources will generate revenue to meet all developmental requirements. Optimism of generating revenue produces near total neglect of other facets of the economy. This neglect is more evident in the agricultural sector. What has emerged is that agriculture is the sacrificial victim of growing our oil sector, neglected in opulent pursuit of petrodollar. The result of this neglect oil's magnified image and pervasiveness in Nigerian economy. So, it is trite to say, agriculture was sacrificed in other for the oil sector to grow. This was not the case in the past. Before Nigeria's independence, different regions were in a sort of open competition over growing different crops. Then, oil was not a determinant of economic viability of a region. The discovery of oil changed the people's perception of agriculture and value system. It seems that Nigeria made a wager on oil and for this reason is now paying a huge price for the inadequate attention paid economic diversification.

Regrettably, the international market is an uneven terrain where food import is also subject to vagaries of oil trade. The rhythmic rise and fall, 'boon and bust' as well as the 'boom and doom' cycles of oil also afflicts the agricultural value chain. Now more than ever, fossil fuel dependent economies are grappling with existential threats at different scales. The global oil shock of 2020 is one of such threats noted to have caused further stress globally. Indisputably, the emergence of a post-petroleum world can drastically change the country economically and politically. Indeed, the post-COVID world could stimulate structural re-engineering within the economy but is it really feasible in a changing terrain where crude oil has been transmogrified into a geomorphic force?

In a stable economy the option is simple. But in an economy where oil is already a cursed resource the option is not very clear. The question emerges whether it is possible to change a cursed resource when it is also a victim not the patient? Proper management of resource offers a clear path to this question. Again this brings us to another issue whether the captains of oil industry will let go of the victim of their economic machination without a fight? Nigeria's mismanagement of her oil wealth has led to social unrest and armed conflicts in the Niger Delta region. Here, regional and ethnic pressures are major determinants of allocation of resources.

Oil's looming crisis is not a far-off chance but a political and economic reality necessitating proper management of the resource. It may seem that petroleum resources are misallocated to private end making the restive population to agitate for equity. This is a time bomb in the hands of the people. Building new bridges for equitable socialization of nature such that intra and inter-temporal equity dimension between present and future are resolved will avert oil's mortality. Unfortunately, the choice is not in the hands of the people. The real decision on allocation of resources rest with the army of rent seekers who sing siren songs of efficiency in the day time but in the night are avid apostles of fossil capitalism.

Unarguably, Nigeria's oil mortality is no longer a far-off chance. A book on this subject alluded to by Professor Shoremeku of Covenant University Ota is no longer waiting to be written. It is being written daily in the minds of every introspective

Nigerian. Indeed, a script about oil mortality is engraved on our national consciousness and this is a diurnal reminder of our vulnerability which is manifested in the form of persistent oil shocks. Eradication of fossil fuel from the global energy economy is now a national reality capable of causing untold suffering- If not properly managed. Today, many Nigerians are apprehensive of the looming danger but are powerless to change the tide of events. Consternation over oil's potential abandonment crisis will not resolve or surmount the intimidating crisis. The inaction from the people is understandable but not taking concrete steps in the right direction by those appointed to surmount rising waves of oil price uncertainties is unpardonable since it tantamount to committing an economic hara-kiri. No nation lives in environmental autarky hence Nigeria must strive to leave within bounds of the global community even when it is detrimental to our short-term economic independence.

A clear instance of swimming against the tide is the recent reforms in the petroleum sector. In an era when most economies are hatching plans to ban fossil fuels production, Nigeria is rebranding in alluring attributes and reinventing the fledging wheels of her petroleum industry. The Petroleum Industry Governance Bill of 2019 serves no other purpose than pandering to private sector appropriation whims neglecting the untold sufferings of the host communities. The reason is farfetched. The simple reason is, Nigeria has an overblown impression of oil's capacity to transform the economy. In this all abiding faith in a single commodity the nation seems to have forgotten the vital lessons of history. From history we learn of how most resources are in flux. In different eons of time different resources were the dominant force of change. The discovery other alternatives changed their dominance. Stone, iron and coal all had their days in the sun. More presently, oil is not void of problems. It is a historical fact that oil is weigh down by different crisis not least of all is its own internal contradictions.

On the realisation of oil's intrinsic deficiencies, most mature and emerging economies have embarked on a carbon neutral route with the hope of not only ending their carbon profligacy but also reducing their carbon upload. One of the countries at the global forefront of this move towards reducing her reliance on oil today is Costa Rica. In 2018, Costa Rica became the first country in the world to ban fossil fuel from its energy supply mix. Others countries like New Zealand followed suit. In November, 2019 New Zealand became the first country to pass a Zero-Carbon Law which sets a net-zero targets for all greenhouse gas emissions by 2050. Since then, several mature economies have joined the league of nations bent on finding other alternatives. The current move away from fossil fuels poses a grave danger to oil exporting nations. It does not only signal the end of oil civilization as it is known today but it will usher in a new dawn- the age of sustainable energy production.

It is pertinent to underscore here that Nigeria's government is not totally oblivious of this happenings. In apprehension of this ominous oil crisis the Vice-President's in the opening address to Oil Producers Trade Section of the Lagos Chamber of Commerce and Industry (LCCI) held on 2nd November, 2017 bade Nigerian energy stakeholders to plan for a future without oil. Despite the public acknowledgement of the impending peril, not much has been done in the form of boots on the ground to place the economy on a path to symbiotic relationship with nature using agricultural transformation as the engine of economic diversification. To a certain extent, Nigeria's faith in fossil fuel is part of the problem hence it is not the solution. Although avenues available to halt this unshakable faith in oil are many but implementation hiatus have stifled realization of these options. In the following section we shall examine the role technology has played in the unfolding reorganization of the global energy economy.

Technological Shift: A Snapshot

Can technology bail the world economy from climate risks in an era when diverse renewable energy solutions have been deployed as Fossil Fuel Exit Strategy (FES)? How can the proliferation of renewable energy sources like wind, solar and geothermal which are the new raves of the moment be made to reduce carbon intensity of the economies without sacrificing economic growth? Is the uptake of Electric Vehicles (EVs) key attribute of this transformation?

Growing interest in technology to bail from prevailing climate risk is not in isolation. Upsurge in climate change together with rising and uncertain cost of oil pushed most oil importing economies into seeking new avenues for curbing huge amount of money committed provision of energy for their people. It is therefore not surprising that these economies that have the prosaic vision of extricating from 'carbon enslavement' are at the global forefront of decarbonization agenda using technology to find feasible alternatives. Correspondently, interest in renewables surged to tally with interest in using readily available resources to address encumbrances to their energy sovereignty.

Electric Vehicle adoption foots the world community's bill. Growing interest in EVs primarily serves the goal of meeting the Paris agreement obligation concurrently meeting their geopolitical interest. Accelerated interest in adoption of EVs is due, in large part, to drastic reduction in cost together with improvements in battery and storage. Critical to the energy transition is the need for efficiency in power generation with proliferation of solar and wind power gaining traction on the basis of their potential of reducing carbon emission in the power generation sector. Reducing energy intensity of the global economy is an outcrop of the Paris Agreement which sets the stage for major fossil fuel-based economies to formulate plans to reduce the carbon intensity of their economies.

With this, a contest between environmental optimists and pessimists who are either avid apostles of renewables or captains of oil industries respectively is on the footing. As one may expect, these diverse camps are battling to keep the flames of their ecological norms alive. Stiff contest over the ecological soul of the global economy is making fossil-led energy economy face an uncertain future. This is due, in the main, to several mature markets search for an alternative energy source. From the pace and urgency attached to making the global economy energy efficient, one can say a kind of revolution within the energy sector is underway. However, proponents of these transformations cannot say with high degree of confidence how this revolution will end. In fact, all those clamouring for end of oil age cannot say with any degree of confidence the outcome of this quest. One thing, however, is certain: it will reduce upload of carbon dioxide.

For one thing, uncertainty surrounding finding alternative energy sources is gradually waning because people are now locked into their different camps. For another thing, it has decentralized energy supply mix and caused further accelerated demand for other renewables. Thus, it can be said that decarbonization measures are pushing the world toward innovations in low carbon technologies. As diverse technologies are deployed, new avenues of creating wealth and jobs are coming on stream. Widespread deployment of renewable energy is an upshot of several interrelated issues paramount amongst which is the access to the sun and wind.

Undoubtedly, the sun is readily accessible in most parts of world but it is essentially a public good. As a public good, the consumption of one entity cannot in any way impair another's satisfaction of it. The concept of public good presupposes that there is non-rivalry in consumption and non-excludability of all parties. The non-rivalry means consumption of a good or service of one party does not diminish the availability of the goods or services to others. Whereas non-excludability, on the other hand, means for technical or cost reasons, it is not feasible to exclude consumers from using the goods and services once it has been produced.

Ralph Nader captures the real reason behind the recent traction of renewables. He puts it thus: 'The use of solar energy has been opened up because the oil industry does not own the sun'.[1] It is pertinent to underscore that the sun is a cheap source of energy. When new technologies are applied to the sun it is turned into energy. It is cheap because it is a technology-driven source of energy, whereas fossil fuel-based source of energy such as petroleum is a commodity subject to market volatility. The cost of technology could drop over time but a commodity like oil cannot be determined because it is subject to market forces. Since new technologies are deployed daily to reduce the cost component of the renewable energy sector, cost of solar power systems globally has fallen by 75% since 2000, while those of energy storage dropped by 60% since 2005 alone[2] whereas fossil fuels' cost is rising.

There is a change in the cost of solar power falling drastically as utility-scale solar PV is procured for about US$80/MWh.[3] The decline also corresponds with natural gas prices in the range of US$7–10 mmbtu—still higher than the US$2–3/mmbtu seen in the shale-rich US in early 2015, but lower than the US$9–10/mmbtu prevailing in Germany and US$14–15 mmbtu in Japan.[4] The result, according to a report by the International Energy Agency (IEA) is the sun could be the world's largest source of electricity by 2050.[5] Due to solar energy's relative cost, the world community invested a total of $2.9 trillion in renewables since 2004 with China leading the pack of investors. For example, in 2017 China invested the largest amount of $126.6 billion, while 157 giga watts (GW) of renewable power were commissioned in the same year globally. (A *Megawatt* is simply 1,000,000 watts of electricity or 1,000 kilowatts of power). Note worthily, wind is at the global forefront of new source of power generation leading the charge towards exit from fossil fuel.

The Global Wind Energy Council (GWEC) reports that the wind sector alone added 52 GW of wind power in 2017 to the global energy generating capacity representing a decrease from 54.6 GW installed in 2015 returning to a 'dramatic growth' in 2019 with cumulative capacity set to reach 840 GW in 2022.[6] The rising profile of renewables in the global energy mix is not music to the ears of fossil fuel optimists for the mere reason that is a doomsayers' gospel recited by those who are wrestling the mantle of energy hegemony from oil producing countries. Beset with rising energy uncertainties, oil magnets are determined not to go out without a fight hence are all out to defend the main source of their enormous wealth. In order to suppress or totally stamp out the looming danger, oil magnets have deployed every known pyrotechnics within their reach to ensure fossil fuels' technological superiority.

Research from fossil fuel citadels is known to have boosted the drive towards upholding fossil fuels which contradicted those carried out from energy efficiency advocates. For example, research from fossil-fuel funds in the USA dangled oil's abundance not the resource eco-friendly propensity or efficiency as its selling point. It is therefore not surprising that a research institution within this fold came out with the consequential finding that it will take about 1.76 million years for present stock of fossil fuels to be exhausted. Whereas Shell BP Energy Outlook of 2015 found that frequent rise in the price of oil is a pointer of its economic viability and relevance in the international political economy.[7]

What these contradictions from sponsored research indicate is that strong forces within the different viewpoints exist. This is an equally linked debate on volatility of the oil market and carbonization of the atmosphere or what is known in some quarters as the weaponization of nature. This debate revolves around whether market forces are sufficient in determining the most efficient source of energy that is not injurious to the wellbeing of the planet. Michael Liebreich of the Bloomberg thinks that "the story should not be how the falling oil prices will affect the shift to clean energy, it should be how the shift to clean energy is impacting on the oil price".[8] DNV GL in their Energy Transition Outlook (ETO) for 2018 (This researcher was part of the global collaboration network for 2018 and 2019 ETO) re-emphasize inevitability of the energy transition forecasting the world will experience rapid energy transition boosted by a strong growth in wind and solar power generation (DNV GL is an international accredited Registrar and classification society headquartered in Hovik, Norway).

The next three decades, according to the DNV GL ETO (2019), will witness rapid decarbonization leading to a decline in coal, oil, and gas in that order. More: coal use already peaked while oil will peak by 2030 but gas will remain the best single source of energy for the world until 2050. The DNV GL ETO (2019) further states: "energy demand will plateau after 2030 mainly owing to efficiencies in the generation and use of energy- even as the world makes steady progress with UN Sustainable Development Goal (SDG) #7."[9] However, growth in the fossil fuels sector would totally cease in 2050 due, in large part, to economic expediency and not so much from threat of climate change we are made to believe since the major determinant of energy transition will be cost.

Limiting emissions from burning of fossil fuels is, therefore, a global emergency requiring urgent attention with the prospects of bringing about comprehensive transformation of the economy at national and international levels. Going by the current trend, fossil energy assets are at risk of becoming abandoned assets while still on the ground given that the value of these assets stated in records will be far too high. Certainly, fossil assets are increasingly becoming abandoned assets and this fact owes to their present abandonment risk status. Bloomberg shades light predicting: "investors in carbon-intensive business could see $6 trillion wasted because policies limiting global warming would stop them from exploiting their coal, oil and gas reserves."[10] This is already happening with the General Electric (GE) recording a $193 billion loss in June, 2019 in fossil fuels investments, amounting to a 74% of its market capitalization lost between 2016 and 2018—so says the Institute for Energy Economics and Financial Analysis.

Effectively, the GE has made a multi-billion-dollar mistake. Some argue the company is suffering from lack of foresight in investment in renewables and bad timing. Others like Stephen Tusa who is the CEO of JP Morgan are of the opinion GE is afflicted with what is termed a 'secular decline'. This implies that GE is threatened by long-term market trends hence power generation business in the years to come could be worse, not better than today. The fate of GE is a cautionary tale for other investors keen on pursuing fossil fuels investment despite growing take-up of renewables globally. Evidently, pledges to double down on climate change are taking its toll. And this major global player (GE) is the first sacrificial victim who is most likely to miss out on the world's shifting use of energy.

All this should be a wakeup call to companies planning to invest in fossil fuel businesses of the associated risk and imminent losses such investment could face. Despite all these indicators, the largest petrochemical industry (Africa's biggest refinery and world largest single-train facility) worth $15 billion is underway in Nigeria- a private sector driven initiative. Such massive investment when the world is trying to divest from fossil fuel requires a second thought. In the face of decarbonisation and transition from fossil fuel into cleaner and other renewable energy sources the project deserves further appraisal. If this huge investment (though a private sector led initiative) is in the renewable energy sector, Nigeria would become one of the leading global actors in this sector. Hence is amongst countries leading pack in transition to a low carbon economy thus improving the adaptive capacity of the people to the perilous effects of climate change. Interestingly, most companies in developed economies have come into terms with the wind of change. Today, they are embracing low carbon technologies while also divesting from fossil fuels assets.

Unmistakably, abandoning fossil fuels either in the ground or reducing uptake of new infrastructure will serve to realize the 1.5 °C global temperature threshold of the Paris Agreement. This, again, is a tricky issue. It is a tricky business because to achieve this global temperature threshold, fossil fuels asset should remain in the ground because they are not only expensive but could impose additional climate-related risk.[11] This is an option most developing economies might not readily latch on to because it might penalise them for externality which they insignificantly contributed to. This is against the backdrop that most developed economies travelled on this pollution intensive route to get their economies to the stage of development they now enjoy. Leaving fossil fuels in the ground unexploited is the elephant in the Paris Agreement's room.

Oil: Black Gold or Devil's Excrement?

Oil evokes different images within the ethical lens of varied individuals, communities, countries and continents. Why is the resource derogatorily known as the 'devil's excrement' and alluring called 'the black gold'? Has the commodity any magical power to change the socioeconomic development of a state? Can the commodity with such mixed attribute be a source of blessing? What are other countries doing to ensure that the curse placed on the resource does not afflict their economies?

In some societies where the discovery of oil is in commercial quantity, the resource enjoys the fortunate notoriety of being a blessing, a source of boom and a means towards immense economic fortune with cities within oil producing regions considered the new *El dorado*. However, in other societies oil is an unfortunate resource, cursed to unleash various forms of socioeconomic impediments on the economy. Oil to such societies is perceived to have a 'curse' cast on it. In such clime, oil is the source of the cyclical distortions and economic shocks witnessed (crude oil refers to petroleum occurring naturally still in pristine form from oil well, or after extraneous substances have been removed. It is hydrocarbon refined into diesel, gasoline, heating oil, jet fuel, kerosene and other petrochemical products).

Interestingly, when oil is a source of blessing, there is a tendency to call it the 'black gold'. In such a society, oil is a blessing because the commodity has magical power of transforming all within its reach (the concept of black gold is deeply connected to the culture of energy sovereignty synonymous with energy security in a stable economy. Crude oil is the black gold because of its colour when extracted from the ground. Oil likened the precious metal gold due to its significant contribution to the wealth of everyone involved in its extraction). When oil is a source of transformational change, oil is the new elixir alluring to everyone wherever its magical spell subsists.

Despite this economic attribute, oil is equally a major source of pollution. It is an incontrovertible fact that this source of energy capable of bringing economic boom is also promoting significant economic doom. Indeed, convergence and divergence cohabitate in one commodity. On this ground, oil production and consumption cannot be without conflicts. Unarguably, oil production exists on a treacherous ground infested with different foibles prone to severe ambiguity. Some of these doubts are function of economic spikes while others are fallout of bio-geophysical complexities associated with extraction and consumption processes. For this reason, crude oil is a veritable commodity capable of inciting diverse emotions within oil producing and oil importing nations alike. Based on its economic value, crude oil is a highly sought resource.

Given its transformational value, most oil producing nations anchored the boat of their economic emancipation on fossil fuel extraction in the hope that it will cushion the economy from short-term economic challenges. The simple reasoning here is discovery of oil would improve the living standards of the people at the same time provide society with other resources to ameliorate long-term risks it has created. With this unshakable faith in oil's power of stimulating economic emancipation, exponential growth was recorded in economic and demographic turfs in some oil producing nations.

Nevertheless, the landscape of oil extraction is not totally free of crisis. One of the negative consequences of massive extraction of crude oil is bifurcation of society causing consequential marginalization of some people. Unquestionably, this is the case in some oil producing countries. The point is worth emphasizing. In most societies oil's magical powers manage to survive, there is an antithetical contradiction-a paradox of stupendous wealth for a select few whilst there is also abject poverty for most of the people. Remarkably, a large proportion of oil producing states face multifarious crisis. In some quarters, arguments exist that oil producing nations are suffering from what is metaphorically termed the 'paradox of plenty,' or 'King Midas problem'.

Generally, these are reference terms and metaphors used to describe countries blessed with large deposits of a scarce resource such as diamond, gold or crude oil. In this paradoxical situation, a resource is maximally exploited through overproduction and overconsumption; such exploitation cannot transform the living standards of the people.

In any economy a large dose of the 'paradox of plenty' persists, oil is a 'devil's excrement' considered a cursed resource symptomatic of what is wrong with the economy. Accordingly, oil is not a source of blessing to such a society but is the devil's excrement (Oil derogatorily is call the devil's excrement for the mystical power of transforming and transformed either for good or for ill). Whilst in societies oil is a source of blessing, it is the 'black gold' viewed a major tool for promoting economic growth. Some may argue oil is primarily a cursed resource because wherever oil is discovered in large quantities, catastrophe usually strikes taking its toll on both local communities and the environment alike.

Expanding this argument, countries dependent on oil as their source of revenue are among the most economically troubled, authoritarian, and conflict-ridden in the world. This is arguable. Terry Lyn Karl came out with further insights on this. Karl's important finding is that countries extensively reliant on exports of "point source" natural resources (meaning those extracted from a narrow geographic or economic base like oil or minerals) are more strongly associated with slower growth. What this objectifies is, mineral-driven, resource-rich countries are amongst the weakest growth performers, despite their high investment and import capacity.[12] Writ large, fossil fuels was material to the industrial revolution noted to have brought about significant improvements in living conditions. It is equally associated with slow growth. On the balance, this sort of generalization is unfounded given oil's immense role in transforming some countries.

Not only oil's intrinsic deficiency that we have learnt from history. Given its critical role of transforming other economies one can agree that oil played a vital function of transforming the world. Within academic circles, various studies point to how millions of people have been taken out of dire economic circumstance with the discovery of crude oil in these societies. To those who hold this 'Cornucopian' view of oil, it is a God-sent resource meant to create wealth hence is a bridge commodity for reducing the poverty gap within the economy and between economies.

Despite this, oil also has a sordid past and a checkered history. In Venezuela, for instance, the 'black gold' evokes an inopportune image of a cursed resource noted to have distorted their national ethical lens in the process causing grave political and socio-economic crises. Today, oil is the source of untold hardship inflicted on the people and a source of malady exacerbating violent conflicts rife within the polity. Oil is not only a curse in Venezuela; the commodity is the root cause of unending civil war in South Sudan.

In another conundrum, Ghana is the latest member of the club of countries whose cultural lens is now blurred due, in the main, to the discovery of crude oil. Transformation of the Ghanaian value system is largely a function of the exacerbation of neo-patrimonial relation which hitherto was non-existent before the discovery of oil in commercial quantity. It is an unfortunate development that the discovery of oil which is supposed to be a source of immense economic blessing inverted the cultural value in the emergence of rent seeking activities. Today, Ghana is experiencing distortions of the country's ethical binocular due, in large part, to rising incidences of pecuniary interest and inequality within the oil industry.

More presently, conflict over who should control the 'black gold' is *belle epoch* of South China Sea crisis—a region waiting to explode over who should control the vast oil wells discovered in the region. China's recent military posturing within the peninsular in July 2020 is with the goal of capturing a large chunk of the oil-laden sea which also serves her geopolitical machination.

Nevertheless, it is not all a sorry sight for oil. Here is why. The good sides of oil have been recorded in history. Interestingly, oil a cursed resource in most African countries is source of enormous blessing for Norway, USA, Saudi Arabia, UAE, Canada, Indonesia and Russia amongst others. This is particularly so because these economies utilized the oil wealth and have immensely benefited from savings they amassed over the years. In these economies oil wealth significantly contributed to amazing transformations of these societies promoting the high standard of living they now enjoy. Be that as it may, oil is not totally free from the blame game of reordering other society's cultural values since it is not void of quandary associated with its extraction. Without doubt, crude oil extraction is pollution intensive. Because of oil's ecological footprint most environmentalists decry the commodity's detrimental pollution for being the source of atmospheric GHG imbalance.

Indeed, oil means different things to different societies and depending on one's ideological leaning the commodity is either a blessing or a curse. Advocates of fossil-led growth who are flag-wavers of the gospel of economic emancipation are still singing siren songs of oil's beneficial attribute of creating jobs and wealth. Conversely, others bemoan this same resource for being the cause of saturation of the atmosphere with GHGs. It is therefore not surprising that the former group prefer oil claiming if they have the opportunity to turn back the hand of the clock on pollution caused by non-renewable resources such as oil, it is doubtful they would have preferred any other option. In this reasoning, turning their back on fossil fuels extraction tantamount to turning away from some of the greatest advancements in modern history.[13]

Arguably, there is merit in the above assumption. Nevertheless, with the world faced with existential threat resulting from this same neo-extractive mentality, it is doubtful if the world can afford to maintain our wanton destruction of the ecosystem without posing new threats to the spaceship Earth. Moreover, the last decade is the warmest in mankind's geological history with 2015, 2016, 2017, 2018 and 2019 the warmest so far recorded, according to the IPCC Special Report of 2019. In the light of this important finding, it is doubtful if the global economy's carbon profligate lifestyle is sustainable in the face of the upsurge of climate risks. With the accelerating rate of carbon uptake, humanity's pollution intensive extraction of oil has fundamentally altered the planet economically and environmentally speaking. The world, unquestionably, is changing and it is changing daily before our eyes!

In recognition of the existential threat, bold steps are being taken to cut down the carbon intensity of the global economy. Reduction in carbon budget of the global economy is causing a structural shift within the energy space thereby altering the landscape of energy consumption and production globally, this much we have already stated. This structural shift makes it inevitable for the world to fundamentally change our ways of doing business as well as the values and norms of socialization of nature, amongst other transformations. Other factors causing these transformations *inter alia,* rapid innovation and declining costs of clean energy technologies; fall in oil prices as an opportunity to advance carbon pricing; fossil fuel subsidy reform; growing international attention to infrastructure investment, particularly in the context of low interest rates; heightened awareness of climate risks in the financial sector; rising interest in low-carbon growth pathways in emerging and developing economies; and an acceleration of the decline in the carbon intensity of the global economy.

It is obvious world leaders are totally committed weaning the global economy from fossil-fuels dependency. What is quite surprising about this is not the efforts made but that most oil producing states are caught napping. In fact, some oil producing nations have not set out in concrete terms an exit strategy for the post-fossil world. Nigeria is one of the countries totally unprepared for this technological shift. Unpreparedness of the Nigerian state is as a result of oil's intrinsic value of creating enormous wealth for the people. With the drum beats of exit heralding imminent danger, this looming crisis could signal the end of Nigerian economic security if proper plans are not laid out. With the global oil glut occurring at the same time with COVID-19 this may signal the total collapse of Nigerian fossil-led economy.

Decarbonization of the global economy is acclaimed the game changer driving the world toward innovative technologies with less carbon footprint. Electric mobility will play an important role in the technological shift. In the following section we shall analyze the electric vehicle and internal combustion engine vehicle debate to understanding how the post-carbon future will emerge and where we stand as a nation.

Electric Vehicles Versus Internal Combustion Engine Vehicles Debate

The debate involving uptake of Electric Vehicles (EVs) and production of Internal Combustion Engine Vehicles (ICEVs) is critical to understanding plans to phase out fossil fuels emissions. This section answers the questions: what is the major gain of Evs or ICEVs and what are the drivers of the proliferation of EVs? Why is it important for the world economy to travel on this new path when an age tested approach exists? Can the EV usher in a dawn of economic and energy security not injurious to economic fortunes of the economies?

In shading light on these questions we examine the debate surrounding ICEVs and EVs with a view to pointing out diverse issues creating the impetus for curbing carbon-intense growth and the role technology most especially eco-efficient technologies play in the emerging world order. This debate revolves around whether limiting fossil fuels should be on moral, economics or saving Mother Earth's grounds. The primary problem here is which of these issues should come first. Arguments exist on whether the decrease in cost of technology should drive change or it is on basis of the morality of using technology with less harmful impact on the people? Other issues include: whether the wellbeing of the planet is far more superior to economic wellbeing of the people who make up the planet? Again, this poses further question of whether benefit of phasing out of fossil fuels is all about the environment alone.

Animated debates at national and international levels have flourished on the above issues targeted at fashioning an equitable blueprint for tackling rising spate of climate risk. Every major debate has various contending camps with dissimilar ecological/economic norms and values raring their heads. On one side are those who believe that humans' high carbon profligate lifestyle deriving from the quest to drive expensive cars inflicted irreparable harm on the planet. With prevalence of climate change, growing need to curb carbon intensity of the global economy in order to deliver the world to a carbon–neutral path are emerging. This will not only reorder the planet's ecological assets but will ultimately usher humankind to an epoch of harmonious relationship with all species of the ecosystem.

On the other side of this debate are those who posit that creating amiable conditions for market forces to proliferate will promote growth hence it should be the main driver of any plan to phase out fossil fuel. In this reasoning, when market forces are unfettered the world will find new technologies to curb imperfections in the system. This group cited innovations in low carbon technologies together with renewables' cost competitiveness and other economic imperatives as the reason for freeing the market in other to rebalance priorities. For instance, the fall in prices of renewables provided an impetus to move towards electric vehicles.

Undeniably, low carbon developments in most cases work for the bottom-line of the customers hence is not merely an issue of it working for nature. No doubt, capital's goal of unfettering the market is to achieve the Triple Bottom-line. The Triple-bottom-line of capital is to serve its purpose not just for profit but serving the people and thus the planet. In this line of reasoning once we have fixed the market, we have also remedied all the Earth's ills which by extension include human and non-human nature alike.

But given the prevailing circumstance and shortcomings of this model, remedying the Earth's ills goes beyond artificial boundaries. It transcends merely reshaping environmental crisis which externalise environmental stress to others. Though market levers can fix a degraded environment it can also be a run towards shortages with the decline in natural resources deposits. For the economy to overcome the shortages, it will require finding an alternative to fossil fuel. This hinges on the fact that as resources become scarce, which by implication is more expensive, all major players in the business arena will have to figure out how to do more with less. In other words, efficiency is a condition scarcity creates not an ecological consideration. So, we can see that efficient utilisation of fewer resources (energy, raw materials water etc.) is the fulcrum of the new drive for renewables.

In continuation of the argument, others point out how saving the Mother Earth from pollution intensive lifestyle will reduce the planet's carbon footprint. Underpinning all these views is an issue of economic nationalism based on the urge to unshackle from fossil fuel cartels' stranglehold on the global energy economy. Effectively, drive to free oil dependent economies from fossil fuel enslavement is the catalyst of green growth not the planetary wellbeing we are make to believe. This underscores the assertion that economic nationalism within oil importing nations plays critical role in determining trends towards finding alternatives given that some countries are bent on breaking up the Organization of Petroleum Exporting Countries (OPEC)'s dominance of the global energy economy.

The current debate as some have argued is a case of cognitive dissonance seeing that some people are either holding on to old beliefs or are anxious because of holding simultaneously contradictory or otherwise incompatible views. (A cognitive dissonance according to psychologists is the mental conflict which occurs when beliefs or assumptions contradicted by new information). The unease or tension such conflict incites is relieved through one of several defensive manoeuvres: they reject, explain away, or avoid the new information or persuade themselves. While it is accurate to say most do care about the environment enough to acknowledge the presence of this pressure, for them it is someone else's problem and not really theirs since they are only interested in its cost component.

Affordability, to this group, is the major determinant of access to a resource not the environmental worries per se. The environmental worries even though present are subject to cost benefits and pocketbook pressures. Some of those who hold this reasoning think it is more about economics than it is about how cares for the environment or health benefits to society. This class of people argue about practical effects on their pocketbook. With the proliferation of EVs, many affordable used (second-hand) plug-in hybrid cars will be readily available soon. Buying an electric car in such a situation is with co-benefits not just to the environment and air quality alone, but also to the pocketbook.

Though this debate is not on the front burner of Nigeria's environmental deliberation, it is relevant to understanding how the post-carbon future will emerge and where we stand as a nation. No doubt, a form of cognitive dissonance has emerged in

Nigeria. Therefore, this treatise will point the country towards what fossil fuel dependent economies are planning and what oil producing nations would be confronted with in the post-COVID-19 future. If the global economy tilts more towards EVs than ICEVs then, global carbon consumption could drastically reduce while high ICEVs production is BAU for oil producers. But with the pace of uptake of EVs, it may seem the world is moving towards the former since low carbon growth with less severe consequences for the environment is gaining traction.

Oil: A Disaster Waiting to Happen?

Is oil a doomed resource driving the world towards an apocalyptic end? Why is the world community pushing for the end of commodity which played an important role of transforming the living standards in societies? Has Oil Age really come to an end? Depending on the side of the ideological divide you belong the answers to the questions will bring consternation or glee.

To the Doomsayers it is a matter of joy an inglorious commodity is meeting its well deserved end. Unsurprisingly, they are heralding imminent end of Oil Age with enthusiasm. Different climate optimists are also echoing the same refrain but with a different optimistic slant of the end of oil. One thing they all have in common is they are announcing the inglorious exit of a seemingly glorious resource. Crude oil is an item of vast historic and economic value demonised and at some point, even vitiated for helping humanity out of past economic deficit. Since the last five years some doomsayers who hitherto were avid apostles of fossil fuels have now become new converts of the gospel of renewables. These new devotees of efficiency who were core fossil fuel capitalists of yesteryears have now put on a new garment of renewables and have become proselytes of the gospel of energy efficiency.

The era of low carbon growth is known variously. While some of the optimists' term it a low carbon revolution; others call it the Third Industrial Revolution; end of Oil Age; fossil fuel exit; carbon–neutral development; post-carbon civilization and clean energy transition. All these nomenclatures have one thing in common: They are principally from the binoculars of different cultures and nations where these characterizations persist. Another thing common to all is the urge to use renewables to curb fossil fuels' stranglehold on the global economy.

Unfortunately, many who were antagonists of fossil fuel have failed to totally exorcise the ghosts of fossil capitalism from their economies. To a large extent they are allowing their trepidations over stranded assets to push them towards taking decisions with long term efficiency gain. Amidst this, the trend towards reducing crippling hold of oil on the global economy will in addition to its short-term environmental benefits reinforce inequalities. It is pertinent to point out that in ascribing exacerbation of inequality to low carbon growth we are echoing growing fear technological gap linked adaptive capacity of most developing countries will magnify. There is the fear that low carbon growth will exclude a vital sector of the economy- the poorest of the poor. Unarguably, the way and manner technologies are being deployed does not make provision to stem inequalities since it was never meant to work for those who are reliant on others for new technologies.

This is against the backdrop of the diverse forms of renewable technologies are floating around in various quarters. Even so, they cannot deliver the world to the 1.5 °C global temperature threshold promised at Paris. If they are to deliver the world then they should elevate developing economies out of the energy poverty and clutches of energy-related market failure through readily adaptive technologies then poor nations must have access to technology at an affordable rate. Though price of renewables is falling daily, there is no luck such technologies will be affordable to the poor soon. At best, the drive towards curbing fossil fuel dominance of the global economy will create awareness of overlapping resource 'crunches' the world is already facing. Still, it cannot totally tackle the problem of inequality in energy gap with developed economies.

A look at forecast on productivity within different regions to 2050 will suffice (Fig. 1.2). Figure 1.2 shows how productivity gap between Africa and the rest of world is widening instead of declining despite the take-up renewables.

From the figure below, it is evident that Africa's consumption and production of energy is almost static indicating that the continent's carbon budget has not seriously changed. Besides, demographic pressure will place new demands for energy infrastructure even though Africa's share in the energy supply is low. Ironically, Africa with huge renewable energy potentials is a net importer of renewables and one of the leading exporters of non-renewables which for instance gave rise to an astronomical 87% increase in the extracted minerals from the continent between 1980 and 2008. Given the current unsustainable growth patterns, this concern cannot be wished away with fashioning of a new rulebook on reporting NDCs or statements of intent on sustaining the environment or wiping the global hunger by 2030. Closing

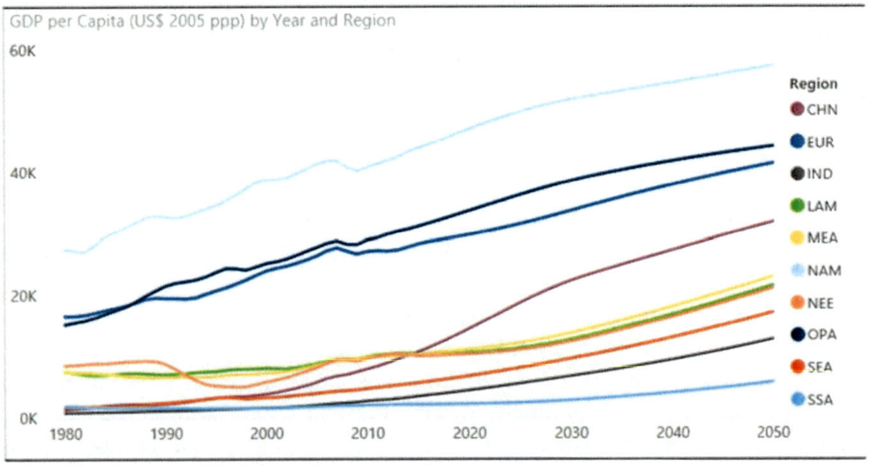

Fig. 1.2 Regional past and future productivity. *Source* ETO (2018)

the energy gap requires concerted global effort capable of engendering an inclusive green growth trajectory. Setting an outline of the trajectory for renewable and non-renewable resource stocks and its likely impacts on countries at different stages of development will be a right step in the right direction. It is a right step because when achieved, green growth agenda agreed upon at Paris would be more evident. Success, here, depends to a large extent on whether it is set within a big picture of what 'non-green growth' means to current and future generations most especially in a monocultural economy like Nigeria.

An Apocalyptic Vision of Oil Age

Predicting petroleum's future is a lively sport within academic, energy, environment and political circles. Assorted pundits are engaged in the onerous task of peeping into the horoscope with the view of prophesying the end of oil Age. Questions such as how the end of Oil Age will occur? When is the likely time? And the effects of the end of Oil Age on the global economy are asked. Different energy outlooks have differing views on this issue instead of resolving the crisis; their divergent opinions are creating further debates on the likely dates consumption of fossil fuels will end. With the debates surrounding this delicate issue, different prognosis on end of fossil fuels have been advanced. None came closest to capturing in objective terms fossil fuels' essence, however. It was Sheik Zaki Yamani, a former Saudi Arabian Petroleum Minister, who prophetically envisioned an apocalyptic vision of the world without oil. At the time, his prophesy (in 2003) was illogical and too farfetched. But considering the unfolding crisis, Yamani has been vindicated given that his apocalyptic vision does ring true. His prophesy on the modality for annulment of this unholy matrimony between fuel fossil and economic growth is quite refreshing. As of late, Yamani words are reverberating within oil producing countries cycles that are now apprehensive of the looming crisis.

Without a doubt, Yamani apocalyptic vision is a diurnal reminder of the consequences of relying on a single resource. More the fact, reining in oil's dominance of energy economy is a testimony of the ingenuity and inherent tendency of humankind to find enduring solutions to the environmental binds the world is enmeshed. In this energy regime, oil producing countries existential crisis is not just from exhaustion of natural resource but from oil's own internal contradictions. It is ironic that the source of modern wealth is creating conditions for its total elimination. It is apposite to underscore that oil lived a chequered life with an equal dose of a fruitful history and a cursed life. In the past, oil rode like a king over the global economy bestowing blessings and curses at will to different societies. Because of this, oil is the dinosaur of modern economy instrumental in creating immense economic benefits; its path is entangled by internal contradictions.

The 'black gold' known to have accelerated economic development which brought about extensive transformation of the living standards of many people is the source of pollution. In this, Nigeria confronts a daunting task. The daunting task Nigeria is entangled today is how to tackle the rising tsunami of renewable energy unleashed by developed economies or perishes for lack of action. As already noted, oil volatility in the international market as well as the commodity's attribute of accelerating climate risk are its major undoing.

At various points in history, different 'big ideas' were promoted to address diverse forms of risks but have had varied consequences. The latest of these 'big ideas' is the deep decarbonization pathway aimed at reducing man-made climate change. Some of those who promoted the deep decarbonization pathway are Jeffrey Sachs and James Hansen. Within this group, there are those who believe fossil fuels should be left on untouched whereas others believe it should be used with moderation with other alternatives developed to limit undue reliance on oil. One of those who are preaching the gospel of decarbonization using renewables is Simon Upton. In his paper titled: 'Implementing the Paris Agreement in world of Fossil Abundance' Upton captures the perplexities within the fossil fuels sector.

This study persuasively argues that there is a fossil fuel disorder in a number of fronts and this disorder is evident in the ongoing confusion within the oil industry's sectoral leaders. Upton in this paper points out how Shell BP in its Energy Outlook of 2018 insists fossil fuel energy would continue to be the dominant form of energy in the near future. In the Outlook, it was further predicted that fossil fuel will power expansion providing around 60% of additional energy. In so doing will account for almost 80% of the total energy supplies in 2035. The most respected International Energy Agency unexpectedly came out with a pro-fossil fuel posture. Whilst a forecast from Chevron casts serious doubts on the "future of fossil fuel and the effectiveness" of carbon pricing as a strategy given that customers want affordable energy.

The Chief Executive Officer of Chevron, John Watson is of the view the world is going to need all forms of energy. For this reason, he claimed fossil-led organizations will remain highly visible in the economic terrain and will be a viable business for a long time to come despite current decarbonization drive. Tony Hayward, the chairman of Glencore, also echoed this view with a similar message for the coal sector. In his remarks to the Paris Business Summit in May 2015 he spiritedly asserts: 'To the fossil fuel incumbents the new normal look a lot like the old one. Hold your hats (if you can); there will be many more holes drilled and mines dug before the curtain goes down on this act'.

This assertion was contradicted by Friends of Earth International who are avid supporters of the exit from fossil fuel (This CSO and other NGOs are at the global forefront of those clamouring to drive down reliance on oil or at best leave it on the ground). According to this group, the unfolding climate crisis is an existential threat, given the world is headed towards CO2 level of 500–600 ppm considered dangerous for human existence. At the current rate of emission the world is headed for 3° (C 5.4 °F) temperature threshold which scientists believe will heighten climate extreme events endangering global food supplies.

Another critical issue propelling the world towards reducing the carbon uptake is the Paris Agreement goal of reaching 1.5 °C global temperature threshold.

Addressing the uptake of carbon via reducing the carbon intensity of the energy infrastructure caused considerable interest in using renewables to bring global emission budget within an acceptable level.

This is not all, however. Several important studies amplify need to reduce the carbon build-up in so doing; reiterating the urgency attached to using alternatives in averting dangerous climate change. According to the 2019 NDC Global Outlook Report, "It's clear business as usual simply isn't good enough anymore. We must do more- much more—in areas related to mitigation, adaptation, and the finance to support all of this work. And we must do it quickly".

With the urgency attached to finding alternatives to fossil fuels, proliferation of solar and wind energy technologies accelerated while development of Electric Vehicles (EVs) is scaled up to cash in on the void exit from fossil fuel could cause. This much we have already stated. In line with the climate emergency leading to proliferation of renewable alternatives, Navigant Research are of the view more than 37 million plug-in Electric Vehicles are expected to be in use by 2025.[14] A research conducted by Bloomberg amplified this noting how electric vehicle revolution is not all about the quantity of cars on the roads.

In real terms, it is more of the reduction of existential threats with cost as another of its co-benefit.[15] In line with this, Bloomberg forecasts that EVs prices will keep dropping after 2025 until initial cost is lower than gasoline or diesel cars.[16] The Energy UK came to the independent finding of how plug-in cars will be more cost competitive than petroleum-powered cars in 2040. But fossil fuel proponents claim that all these are mere sound bites from 'tree huggers' and 'naysayers'. Given these divergent views, we are in an era of a 'free for all' fight between two diametrically opposed forces within the energy sector.

One of the umpires of this zero-sum game between diametrically opposed groups in the climate endgame is IPCC. According to the IPCC, a deep decarbonisation to below 2 °C is meant not only to end undesirable GHG emissions but will strengthen sustainable development principles. While some believe such assumption is flawed because elimination of fossil fuels is not feasible within the prevailing economic circumstance. Instead, quest to curb fossil fuel dependence will foster institutionalisation of profit and hegemony of the economic worldview. Whereas others decry ongoing decarbonization claiming it strives to impose a green economy on capital poor countries together with it is associated shortage crisis. Although some concede the world will in the short-run experience massive scarcity but eventually growth in the green economy will improve the quality of life. But the gains derived from greening the economy will dissipate when confronted with scarcity crisis which will further reinforce anthropogenic climate change.

Taking this argument further, limiting fossil fuels consumption is an option developed economies quickly lashed onto given their level of development. In developed economies, putting an end to the Oil Age is a viable option if it can limit carbon emission. Resolving crisis involving the quest to eliminate fossil fuel from the global energy mix is entirely a different matter in developing economies. In Africa and the rest of the developing world, Oil Age cannot really end without first addressing some complex issues underlining oil production and consumption. What one can say now is, in the haste to replace the fossil world order with a new one, the world leaders seem to have forgotten the vital lessons of history. From history, we learnt the end of Stone Age had very little to do with shortage of stone crisis. In a similar manner, transformation from the steam engine had little to do with limits to coal production. Hence, the renewable energy revolution will have less to do with scarcity or fear of oil running out. Quite to the contrary, fossil dependent era will end not from resources limits. Humanity's fossil dependency will end when there is realisation in addition to limits to economization of nature, there are other barriers which nature can impose.

Because of the world's high carbon consumption, we have just over a decade to get climate emergency under control, so say U.N. top climate scientists in 2019. In the 2018 and 2019 Special Report of IPCC on 1.5 ºC and that on Climate Change and Land, top scientists re-emphasized urgency attached to curbing the adverse carbon footprint of fossil fuels from the global energy mix. Note worthily, most scientific papers agree that the year 2030 is the likely date the world should totally decouple from carbon-led development towards an energy mix skewed towards renewable alternatives. The IPCC Report further finds that near-term risk will magnify despite current efforts at decarbonisation with the world projected to record more devastations in the coming years. On this ground, unprecedented change in all aspects of life is needed if we are to reduce global warming to 1.5 ºC threshold. One of the key messages emerging from the report is we are already witnessing consequences of 1 °C of global warming through increases in extreme weather, rising sea levels and diminishing Arctic sea ice, among other changes.

It is trite to say that the world stands on the brink of disaster when it comes to holding global warming to moderate levels, and varied nations need to take unprecedented actions to cut their carbon emissions. In the next decade (precisely by 2030) radical changes to bring the global mean temperature to within 2 °C threshold considered safe for human existence are required. With pronouncements of President Donald Trump of the USA on climate change as China's hoax and commencement of the formal withdrawal process in November 2019 from the Paris Climate Agreement, meeting this ambitious goal will be a formidable and daunting task requiring all parties to the Agreement to significantly step up action and raise their NDC ambition during the 2020 review process. Not just because America is the world second largest emitter of CO_2 but more of the fact that the country is a major player at the international level with most nations following their lead (Kyoto Protocol failed largely because of America's intransigence. Similar problem is now trailing the Paris Agreement).

There is the imperative for radical transformation of human civilization at a magnitude never recorded in man's history. Such transformation calls for sweeping reductions in emissions in the next decade from all emitters. This would probably need to be more than 1 billion tons per year, larger than the current emissions of all but a few of the very large emitting countries. Various studies point to the fact that in 2050 a total or near-total phase-out of the burning of coal will stave off further degeneration of the global temperature. Emission budget of advanced economies projected to rise (Fig. 1.3) increasing the risk of climate change.

The bottom-line emission (Fig. 1.1 above emission scenario) from major global economies shows the world is off target to meet Paris commitments. To avert dangerous climate change, the world will need to develop large-scale "negative emissions" programs to remove significant volumes of carbon dioxide from the atmosphere. Although the basic technologies exist, they have not caught on widely, and the IPCC Special Report of 2018 has strongly questioned whether such a program can be scaled up in the brief period available.

This report was unanimously unacceptable at the meeting held to consider the report because major oil producing nations such as Saudi Arabia were all out to defend their economic interest. They decried alarmism of the IPCC whilst optimists claim the IPCC language toned down gravity of the impending environmental disaster waiting to happen.

This controversy trailed the final report making it a trade-off between different forces. As Phil Duffy, President of Woods Hole Research Centre puts it: "If you're

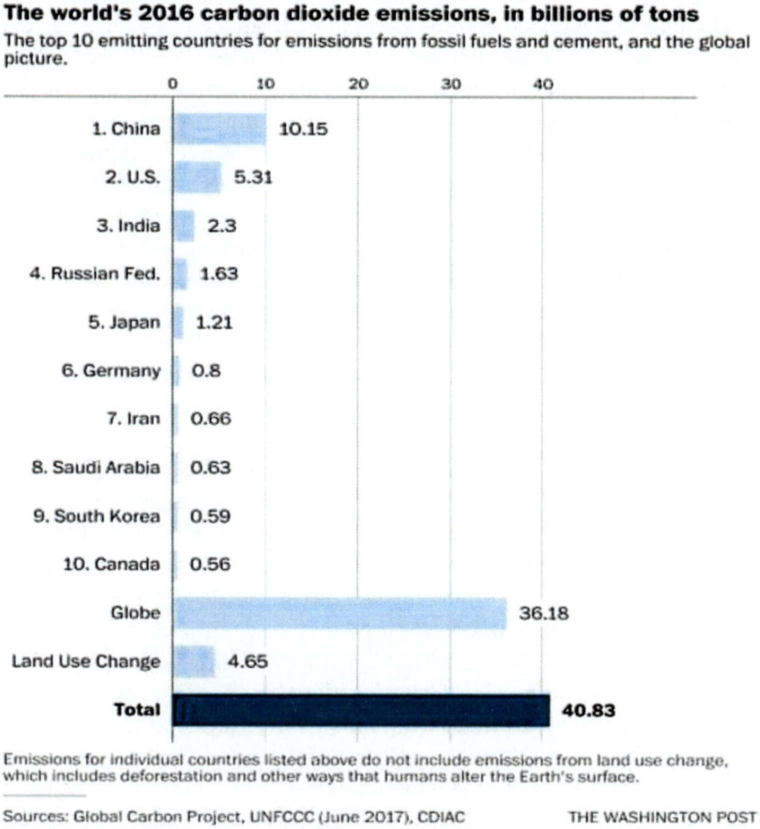

Fig. 1.3 World's carbon dioxide emission in 2016. *Source* Global Carbon Project (UNFCCC, 2017)

expecting IPCC to jump up and down and wave red flags, you're going to be disappointed. They're going to do what they always do, which is to release very cautious reports in the extremely dispassionate language." Whilst Oliver Gedan, head of the Research Division of the German Institute for International and Security Affairs believes the new approach buys some time and "resets the clock for 1.5 degrees Celsius to 'five minutes to midnight".[17]

In Chris Moonley and Bradley Dennis opinion holding the warming to 1.5 degrees Celsius could save an Alaska-size area of the Arctic from permafrost thaw, muting a feedback loop that could lead to still more global emissions.[18] To avoid that in barely more than 10 years (from 2020), the world's percentage of electricity from renewables such as solar and wind power would have to jump from current 24% to something more like 50 or 60%. Coal and gas plants in operation would be equipped with technologies not tested thus far, collectively (such as Carbon Capture and Storage (CCS), Bio Energy Carbon Capture and Storage (BECCS) and Solar Radiation Management) capable of preventing plants from emitting carbon dioxide into the air by funnelling it to be buried underground as well as reducing impact of the sun warming the earth surface. The duo's conclusion is most coal plants would shut down while cars and other forms of transportation would need to shift strongly toward being electrified; powered by renewable energy source before 2050.

The DNV GL ETO (2018) Outlook also holds a similar view predicting energy related emissions will have to stay flat over the next decade, reaching the highest point in 2025, some 3% higher than today. Based on DNV GL ETO forecasts of primary energy use, by 2050 (Fig. 1.2) emissions from coal will have declined by almost 65%, emissions from oil by 52%, but related emissions will have increased by 6% with only sub-Saharan Africa the only region continuously increasing emission from a very low base. Viewed from the other end of the telescope (and from the personal experience of this researcher and as part of the collaboration network who contributed to the DNV GL ETO 2018 Outlook, Africa's emission gap is largely fallout of the low carbon uptake resulting from an equally low economic base) the carbon intensity of African economies is due to the less than 1 $GtCO_2$carbon budget forecast in 2050 and not any conscious effort at decarbonisation per se. The DNV GL ETO (2018) forecast shows an even more dynamic transition on the supply side of the equation as electrification of industry and society accelerates towards 2050 (Fig. 1.4. To satisfy 200 EJ of electricity per person demand in 2050, electricity generation will be 240 EJ,—with the apparent 18% surplus coming from losses in transmission and distribution and power used by the sector itself.[19]

Key results from the DNV GL ETO (2018) model shows demand will peak in 2032 and then slowly decline. This means the global primary energy supply required to satisfy demand mirrors this trend within their forecast period. Although demand drops only by 10 EJ per year from the peak in 2050, supply drops 50 EJ per year due to rising energy efficiency in power generation.

Undoubtedly, energy efficiency would accelerate energy transition in the years to come. From DNV GL ETO (2018) prediction, the world's energy system is highly sensitive to changes in energy efficiency. For this reason, the world's energy intensity—units of energy per unit of GDP—has been declining on average by 1.4% per

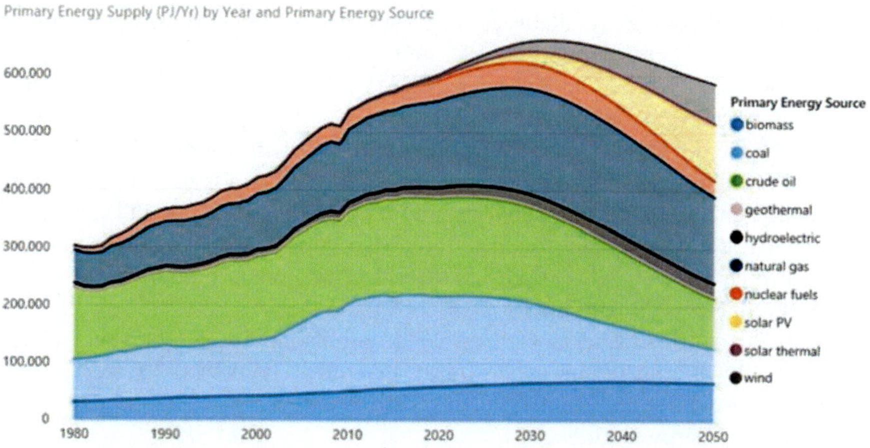

Fig. 1.4 Projections of primary energy source to 2050. *Source* DVN GL, ETO[19]

year for the last two decades. In DNV GL ETO (2018) forecast, this will almost double to an average annual decrease of 2.5% resulting from the accelerating electrification of the energy system. Essentially, using electricity rather than fossil fuels is much more efficient, with less heat losses. DNV GL ETO (2018) claims that the ramp-up rates of energy efficiency are dependent not only on new combustion systems, battery developments, and other engineering innovations like 3D printing with automation and digitization as key enablers of improvements in manufacturing processes in the design and operation of buildings.

The IPCC Special Report (2018) on the need to ditch coal and other fossil fuels from the energy supply mix received serious criticism from the World Coal Association. According to The Washington Post of 7th October, 2018, Chief Executive, Kate Warrick in reacting to the reports notes: "While we are still reviewing the draft (IPCC Special Report), the World Coal Association observes that any credible pathway to meeting the 1.5-degree scenario must focus on emissions rather than fuel, which is why CCS is so vital."[17] The attempt to water down the strong wordings of the Special Report is not limited to the World Coal Association, however. It was also reported by the newspaper that Saudi Arabia, the oil giant attempted blocked the adoption of the UN climate change report unless a passage during the UNFCCC COP24 in Poland highlighting limitation of the national carbon-cutting pledges were either removed or altered. A major reason for their opposition to IPCC wordings is that most of the scenarios projected to reduce the carbon intensity of the global economy involve sharp reduction in the use of fossil fuels.

Notes

1. Ralph Nader, quoted in Linda Botts, ed.., Loose talk, 1980.
2. Van der Hoeven, M., 2015. Opportunity to act: Making smart decisions in a time of low oil prices. International Energy Agency presentation at the Oxford

Energy Colloquium. Available at: https://www.iea.org/newsroomandevents/speeches/150127_OxfordEnergyColloquiumspeech.pdf.

3. Quandl, n.d. Natural Gas Prices and Charts. Available at: https://www.quandl.com/c/markets/natural-gas.
4. (MMBTU, or MBTU, stands for one million British Thermal Units (BTU). A BTU is a measure of the energy content in fuel, and is used in the power, steam generation, heating and air conditioning industries. One BTU is equivalent to1.06 J. MM" stands for one million - it originates with the Roman numeral M, which stands for 1,000. MM is therefore "one thousand thousand," or one million). New Climate Report 2015.
5. IEA, 2014. *World Energy Outlook 2014*. International Energy Agency, Paris. Available at: https://www.oecd-ilibrary.org/content/book/weo-2014-en.
6. Global Wind Energy Council (GWEC) 2017.
7. Shell BP Energy Outlook, 2015.
8. Liebreich, M., 2015. State of the Industry Keynote. Presented at the Bloomberg NewEnergy Finance Annual Summit, New York, 14 April. Available at: https://about.bnef.com/presentations/liebreich-state-industry-keynote/.
9. DNV GL Energy Transition Outlook, 2017 https://eto.dnvgl.com/2017/.
10. Bloombergy[BNEF], (2017).
11. Hansen, J. (2012a) "Game Over for the Climate," *New York Times*, May 9, 2012, available from https://nytimes.com.
12. Karl, T. N. (1997) The Paradox of Plenty: Oil booms and Petro-States: Berkeley: University of California, 1997. p. 342.
13. Dr. Charles Donovan (2017) Climate change: what investors need to know. In this article Dr *Charles Donovan.* 16 February *2017*. Dr *Charles Donovan*, Director *of* the Centre for Climate Finance and Investment at Imperial College Business School, warns investors to prepare for the effects *of* exit from fossil fuel, International Monetary Fund, the direct subsidy to *fossil fuels* globally is nearly three times that to renewable energy Available at: https://www.google.com/search?q=+Charles+Donovan+2017+o+exit+from+fossil+fuels&ie=utf-8&oe=utf-8&client=firefox-b.
14. Navigant Research (2017) [Online] available from <https://www.navigantresearch.com/newsroom/more-than-37-million–plug-in-electric-vehicles-are-expected-to-be-in-use-in-2025> [12/12/2018].
15. Bloomberg New Energy Finance [BNEF] (2017) Available from <https://www.bloomberg.com/news/articles/2017/07/06/the-electric-car-revolution-is-accelerating> [12/12/2018].
16. Ibid.
17. The Washington Post of 7th october, 2018 available at: https://www.washingtonpost.com/amphtml/energy-environment/2018/10/08/world-has-only-years-get-climate-change-under-controlunscientists-say/?noredirect=on).
18. The Washington Post of 7th October, 2018.
19. DNV GL ETO 2018 Outlook (https://eto.dnvglcom/2017/main-report.

Chapter 2
COVID-19 and Nigeria

> In every crisis, doubt or confusion, take the higher path- the path of compassion, courage, understanding and love.
> Amit Ray, *Nonviolence the Transforming Power*
> 'These so-called bleak times are necessary to go to a much, much better place.'
> David lynch

Origin of Coronavirus

Where did COVID-19 originate from and how prepared is Nigeria to confront COVID-19 impact on the people? What is the economic cost of coronavirus? Can Nigeria look inwards for local solutions to guarantee medical supplies, food security, high quality and affordable healthcare? What is the nexus of COVID-19 with LCD and why is the economic recovery plans tied to fashioning a carbon neutral path? These are the questions this chapter shall answer.

The coronavirus disease of 2019 (COVID-19) is a highly transmittable and pathogenic viral infection caused by severe acute respiratory syndrome coronavirus 2 (SARS-CoV-2), which emerged in Wuhan, China and spread around the world. Coronavirus came into global attention in Wuhan city of Hubei province in December 2019. The source of the virus is not known, it is alleged that wild animal species like bats, snakes, birds, rabbits and marmots are bearers of the coronavirus but no one knows how it was initially transmitted to humans. Human to human transmission is what resulted into community spread which has now become predominant. This spread results from close contact with carriers of the virus who are not aware they are infected with the virus. In May 2020, a total of 4.44 million cases were reported across 188 countries resulting in more than 300,000 deaths while a total of 1.58 million recovered from the disease. There is no vaccine for the pandemic at the moment. Several studies aimed at producing its vaccine are ongoing.

COVID-19-relaled economic meltdown is now the new normal yet, the pandemic is not just a public health issue. It is also an economic and social problem. Disruptions of air, land and sea transport systems within and between countries due to lockdown

A. S. Okoh, *Oil Mortality in Post-Fossil Fuel Era Nigeria*,
https://doi.org/10.1007/978-3-030-60785-2_2

initiated by different countries compounded the situation. Of note, disruptions at national and international levels led to the multilateral actions in the pre-COVID-19 era. Equally significant, the temporary ban imposed on exporters of essential commodities also gave birth to unilateral actions. Consequently, unilateral moments received renewed attention with the impetus derived from inter and intra state nature of the pandemic. According to the World Customs Organization reported, a total of 32 countries and territories, adopted stringent and immediate export restrictions on critical medical supplies and drugs that were specifically meant to respond to COVID-19. Around the world different leaders announced export restrictions on agriculture produce in preparation for food scarcity foisted by COVID-19.

Economic Implications for Nigeria

The year 2020 is a year of multiple tragedies. It is year when COVID-19 unleashed severe devastation on the world. This year is also when a combination of issues including COVID-19 pandemic accelerated the downward spiral of the oil market making it not just a public health challenge but also an accelerant of food crisis. The severity of COVID-19 makes it not only a time of uncertainties but is also a time when certainties are dangerous. Overcoming uncertainties of an abnormal time requires an abnormal approach. According to a United Nations Economic Commission for Africa (UNECA) report, Nigeria's loss from oil revenue alone in the year 2020 as a result of the pandemic would be between US$14 billion and US$19.2 billion.

McKinsey & Co are of the opinion that the coronavirus puts 150 Million Africa Jobs at Risk.[2] That, there will be between 30 to 35 million reductions in wages and work hours occasioned by the pandemic while in manufacturing, tourism, wholesale and retail sectors will witness more than half of the workforce job risks as the continent will require US$5 billion in 100 days in the face of COVID-19 pandemic. Most African countries Nigeria inclusive were unprepared for these risks hence severely affected because of their undue reliance on food imports. Shortage in rice import is outcome of exporting countries decision to reduce exports to protect local markets from imminent food scarcity, Africa and the rest of the externally sourced food dependent nations cannot import supplies needed. Today, many of these countries are grappling with a bleak future because severe famine will ravage many communities resulting from the dislocations of agricultural and economic activities within and between different countries. Last part of the existential risk is climate change known to have been ravaging many communities on a yearly basis.

The pandemic disrupted economic and agricultural activities creating an economic emergency. But the devastation of the pandemic was not only in the economic terrain since it also created for food shortages (Fig. 2.1).

In Nigerian Bureau of Statistic's projection there will be a 5% drop in growth in 2020 with unemployment rising to at least 40%. Nigerian GDP is expected to decline by at least 3% while the May 2020 unemployment rate jumped to a record high.

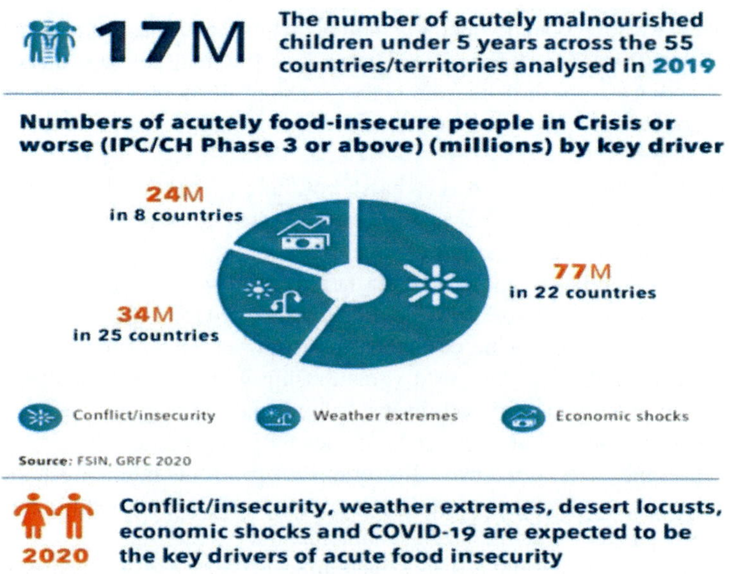

Fig. 2.1 The number of Persons affected by coronavirus and the Conflict/Food insecurity Status

Effect of Coronavirus on the Global Economy

The Coronavirus and slump in the global market are the defining challenges of modern age. In March 2020, exponential increase in cases of COVID-19 casted a spell over many economies. A major downside of this virus is the lockdown of many countries prompting reduced fossil fuel consumption. Together, these threats are pushing many towards severe shocks and stress. With exacerbation of health challenges, Nigeria like the rest of the world is paying a huge price for profiting from obnoxious trade in endangered species. Interestingly, most people blame China's weird desire for endangered species for placing humankind in this dire situation. Instead of blaming China, these individuals should blame themselves as their acquiescence aided and abetted debauchery of nature. Crying wolf now won't stop propagation of the pandemic. No amount of blame game has changed intensification of the coronavirus. If we are to place the blame then we must accept that Africa is not totally free from the charge of creating amiable environment for diffusion of the virus because Africa's socioeconomic and political challenges aggravated the continent's vulnerability.

In the advent of COVID-19, Nigerian economy is on a free fall with economic and human health challenges all occurring concurrently. Arguably, the country is in one big giant merry-go-round roller coaster trip on economic recovery with insignificant interest paid diversifying the source of revenue. Hence it can be said that the country's half-hearted commitment to diversification is to blame for the sufferings recorded across the nation. Though government formulated different health care policies but the coronavirus crisis of 2020 caught the nation napping. For one thing, the health care services are poor. For another thing, policy inconsistency has impaired providing affordable health care for the people. The nations frequent policy somersaults has made the country to be just like the barber's chair, all motion no movement. So, what the country is witnessing in the form of perennial economic recession is a self-inflicted wound.

Even though this may sound very harsh but there is merit in the assertion. This owes, in large part, to displacement of agriculture as the main foreign exchange earner. No harm done in relying on a single product if other sectors such as agriculture grow at the same rate. What is not acceptable here is the overt optimization of the oil sector over agricultural sector. With the urgency attached to finding lasting solutions to the pandemic there is a wind of change in the air but this change. A true change requires transformation of the world's relationship with nature. This also requires a cultural and economic shift both in institutions and the way run the economy. It further entails reappraisal of micro and macroeconomic conditions leading to an exit from our present material intensive lifestyle. Exit from this dilemma should focus growth of Nigerian economy towards eco-friendly relationship with other species based on eco-efficient utilization of available natural resources.

Since infinite growth is not possible in a finite world, we can only achieve ecological security if there is a major shift in destruction of other species where the world's extraction pathology is geared towards a pro-environment and post-growth society. The new society emanating from this should be one where the economic ideology of nature is balanced in a manner that economic processes of competition and optimization decouples from natural capital depletion. This goes beyond current business as usual 'brown' resource intensive development path where a single commodity such as crude oil can bring gloom and boom to the economy. Forging a post-petroleum society should be the goal of such endeavour. A slip out from the mantra of growth: the more is waste; the bigger the common wealth to a more ecologically benign interaction with nature.

This new social contract with nature should use less of market-based instruments for ecosystem protection. Pushing the country's consumption and production towards greener technologies is the responsibility of government. Relying on cultural institution is inevitable if the people must embrace the cultural shift. Such transformation may be tainted by the rejuvenation of post-colonial era relations where animal tourist enclaves prevail alongside pillage of endangered species. A shift in the present architecture will require carefully crafted policies aimed at aligning markets towards environmental resources goals where Climate Smart Agriculture is promoted. Governments should play a strong leadership role by creating individual incentives (including negative ones) to align consumption and production decisions with significant health, economic and environmental consequences. They should create incentives capable of facilitating achievement of the goal of agricultural development while simultaneously bringing about equitable consumption of natural resources.

All said, the global economic meltdown is far from bottoming out, so Nigeria is not totally free from the hazards resulting from the current dysfunctional system. Though market trends are upward bound it is not yet Uhuru. This pandemic can still mutate into a more devastating form. One thing is, however, clear. It is a warning that promoting or encouraging the positive values of the environment should not be taken lightly. This view presupposes that 'doing well by doing good' philosophy, would not turn 'genesis backward'. It seems that the world community in their rush for a 'restorative economy' to bailout the global economy have succeeded in reengineering nature. The mad rush to tame nature has given birth to a hydra-headed monster or what Heather Rogers calls a 'green gone wrong' economy.

Care should be taken so that it does not turn the world community into remnants of history. At the rate the virus is disseminating, it seems the world has not heard the last of COVID-19. Unfortunately, Nigeria is still handling it with a kid's glove. Thus, we should re-examine our travel and border laws. The hasty reopening of the country after the lockdown is ill-motivated. The primary goal of the reopening is for optimization of oil export but care be taken it does not consume the nation. In this the country is bent on materialization to offset budget deficit. Government seems to have placed immediate economic benefit above human health. There should be re-examination of the re-opening of the country for monetary. This also includes examining the process of dematerialization based on super materialization wherein cost is rendered invisible by spatial or temporal displacement. Such orientation gives the lie to the idea that humankind can buy its way out of all sorts of threats.

Government corrective measure is necessary to guide towards greener alternatives while also creating new incentives for capitalization by way of providing part of the initial capital for investment in agriculture. States should provide an amiable environment for the fertilization of ideas to encourage businesses to invest in the agricultural sector. Transformation of the agricultural value chain is a sure path out of the uncertainties of the global energy economy. This requires phasing out of subsidies, reforming policies and providing new incentives, strengthening market infrastructure and market-based mechanisms, redirecting public investment and greening public procurement.

As these policies are rolled out, the private sector in Nigeria should rise to the challenge. They should rise to take advantage of the unfolding technologies by also searching for less expensive technologies which address Nigeria's peculiarities. Progress in the agricultural sector is a sure path to economic sustainability hence government should allow the numerous opportunities embedded in implementing CSA to permeate all facets of the agricultural value chain.

Implications of Fossil-Fuel Exit Strategy and COVID-19 for Nigeria

A country which is on path of becoming the 20th developed economy in the world is suddenly borrowing large sums of money to foot their budgets and COVID-19 recovery? Nigeria is supposed to have attained the lofty goal of becoming the 20th most developed economy in 2020? Why is oil price resurgence in the international market in 2019 inadequate to finance the budgets with government reviewing it budget benchmark from \$57 to \$20? Should the country after every bout of shock go under the water only to come up and still commit the same error over and over? These perplexing questions are crucial to Nigeria's petroleum-led economy. We have discussed the impending implosion of the global energy economy and COVID-19 pandemic to this point but we have not narrowed our discussion to the Nigerian context. We also have looked at different states plans and technology for lifting their people out of near-term risks but have not looked at reasons for Nigeria's lackluster performance after every cycle of oil shocks.

There are many grey areas of Nigeria's socialization of nature and using its vast oil deposits to lift the people out of poverty. As far as Nigeria is concerned the abundant fossil fuel assets considered the 'golden goose' because it has been laying the golden eggs might become a stranded asset in the near future. This may result from near term risk such as COVID-19 pushing the world towards re-appraisal of consumption and production of hydrocarbon and long-term goal to decarbonize and build an inclusive green economy. Arising from this, Nigeria's oil industry could see less demand from existing importers who are either highly invested in renewable infrastructures or weigh down by coronavirus imposed stress.

Though phasing out fossil fuel consumption is a good idea, it is a painful joy. In the near future, Nigeria would be headed for stifling economic conditions because it failed to plan for a period when the oil windfall winds down or a pandemic will impose new conditions on production of oil[3]. This new age is an opportunity for the country to reap enormous economic benefits from elimination of crude oil in the global energy mix if there are sound policies promoting economic diversification. Depending on the side of this argument you are, the dilemma of ending fossil fuel dominance of the global energy mix could lead you smiling to the bank or it may be a road paved with sweats and blood of the people.

Still, it is not totally a sorry sight we are made to believe. Quite the contrary, liberation from oil could be an opportunity to rebalance the degenerating oil industry away from its profligacy using innovative finance to create an economy where there is generation of green jobs through strategies targeted at economic diversification. Furthermore, an exit from fossil fuel could be the cure for the dysfunction between 'resources curse' and economic growth while also fostering geophysical sanity in the politically restive and environmentally unstable oil producing region of the country.

The question Nigeria needs to ask is no longer when the implosion of her energy economy will commence but how prepared are we to run this existential race threatening our corporate entity?[4] How the country utilizes its vast stock of oil in a manner that it is not a negative change maker but instead is an agent of ecological neutrality capable of ensuring economic progress in building a post-carbon civilization is our ultimate goal. The primary goal is, therefore, to address threats Nigeria could face in the plans of curbing fossil fuels from the global energy supply mix. This is with a view to equipping policy implementers with a policy toolkit to overcome the vagaries of climate change simultaneously eliminating uncertainties associated with decline in the country's oil revenue.

The downside of this is maintaining an equitable growth trajectory in the post-COVID era against the backdrop of oil shocks without bringing severe pressures on the fragile economy. Such an endeavor is a tough ask. Managing discontinuities associated with curbing fossil fuel extraction could have serious repercussions for GDP growth. Given that the economy is a monoculture, with crude oil accounting for over 90% of export earnings, shifting to a new commodity with the decarbonization could disrupt the pace of human development. Plans to curb fossil fuel dependency would exert serious pressures on the country, but not taking prompt actions on the carbonization of the ecosphere is far more dangerous for the people.[5]COVID-19 has exposed Nigeria's underbelly. Evidently, it is a weak underbelly where any external shock can expose the country to further stress.

Again, the balance of argument calls for taking prompt actions on the current global agroecological emergency because not taking action will surcharge future generations for present generation's eco-decadent lifestyle. A reason for this assertion is that the environment present generation is enjoying will be far healthier than those future generations would have. Resolving the paradox of cohabiting extremes and antithetical contradiction of pursuing two inversely related paths in the post-Paris/COVID-19 era should be of interest to government. Inconsistencies of revamping Nigeria's oil sector in this new age together with lack of pro-poor carbon agenda for improving human health while diversifying the economy are major challenges in the years to come.

For now, policy attention is on the emergent oil price dynamics and volatility of the international market. It is rather unfortunate that at the height of oil boom the leaders did not plan. And as such are now facing the stark reality of oil price instability together with COVID-19 lockdown. It has become increasingly clear to all discerning observers that all is not well with the Nigerian state- economy wise. The current probe of the NDDC in July 2020 by the NASS is an eye opener that the country is held in the vicious grip of corruption politicians. According to Channel TV report of 20th July, 2020, NDDC spent the humongous amount of 87 billion from February 2020 to July 2020 but did not execute any project. The ensuing charade in the NASS between Ministers and legislators of allegation and counter allegation is not tenable.

Certainly, managing fossil fuel-related crisis resulting from decarbonization and coronavirus disease are complex issue for major oil producing countries. It is a complex issue because these existential crises can undermine the corporate entity, people and the environment. Given the gravity of this crisis, Nigeria could be on the road to perdition if properly crafted policies are not in place. Options available for government are very complex which entails toeing 'deep decarbonisation pathway'; or face an upsurge of climate risk with new exacerbated economic hardship; or the country can proceed with the present 'drill, baby drill' model with expected unpleasant outcome from fumes of poisonous gases experienced in some part of the country, most especially, fumes associated with oil exploration. This has both environmental and health implications for the host community and by extension the country. Whatever alternative Nigeria adopts will have severe consequences. This is the devil's alternative in Nigerian context or Nigeria's carbon catch-22 situation. No doubt, there is a stiff penalty to pay for any of the choices and the earlier the country face up to this reality, the better for all.

The price of Nigeria's ecological indecisiveness, for instance, could be in the form of lost environmental quality and exacerbation of different pandemics thus leading to further destructions of the fragile ecosystem. Whilst the price of pursuing pro-nature values is the decline in pace of economic development resulting from exit from fossil fuel which, in turn, is linked to inability of government to flesh out an equitable blueprint to navigate between the murky waters of ecological and economic uncertainties. Indeed, whatever option Nigeria adopts today, there will be significant consequences.

A major premise for Nigeria's oil dilemma is the country's pollution-intensive economy. Primarily, Nigeria's source of pollution is through conversion process of energy with oil extraction and exploration contributing significant quantities of emissions. Paradoxically, too, crude oil is a major income earner (60%) which significantly contributes to foreign exchange earnings (90%). There is an inverse relationship between energy and growth because energy intensity of the economy is critical to growth. Rise in GDP and improvement in the living standard of the people are mutually relational due to their propensity to escalate carbon emissions. Fighting COVID-19 using these earnings will in the short run improve human health and the living conditions of the people but is linked further to pollution. For Nigeria to[6] grow the economy, it would require high carbon uptake hence is with serious entropic dynamics.

This is against the backdrop that the current energy plan is not just inadequate but lionises capital accumulation in exchange for environmental quality. Not only that it is massively expensive, it has more to do with rent seeking proclivities of the political class who encourage debauchery of both human and non-human nature. Nigerian leaders have the options either to get their acts right or end futile exercise of building new stranded assets. To bring about change, the political class must remedy flaws in how the country views and relates with ecological assets and new emergencies. If deployment of a fossil infrastructure will be ecologically unsustainable, then, the best option is to call the whole thing off and take a path most developed economies are now toeing.

Where Did We Go Wrong?

Can Nigeria really wean the economy off hydrocarbons without sacrificing growth? How can the country pay rising debt in post-COVID-19 era? What are other countries doing right and Nigeria is not?

There are various answers floating around on these questions with different pundits claiming to be harbingers of economic salvation. There are diverse solutions flowing out from informed and uninformed public. Still, the issues remain unsolved in view of the fact that the malady cannot be wished away. To the poor on whose toil Nigeria's economic debt is paid, oil wealth symbolizes the problem with Nigerian economy as it is also the solution. Whereas for the superrich; oil instability is part of the normal business cycle.

To understand what went wrong with oil wealth in Nigeria, it is important we examine what is obtainable in the oil sector of other economies. The discovery of crude oil in Norway was the major source of impressive rise in status of the country, for instance. Norway is one of the most developed economies in the world with a per capita equivalent of 711% of the world average with an average of 59,631.16 USD from 1960 until 2016, reaching an all-time high of 91,594.18 USD in 2017 from a record low of 23,176.96 USD in 1960. Similarly, Indonesia considered Nigeria's development twins transformed her economy using the same petroleum resources considered in Nigeria by some as a 'resource curse'.[7]

Joseph E. Stieglitz in 'Making Natural Resources into a Blessing rather than a Curse' succinctly puts it in more context: 'Some 30 years ago, Indonesia and Nigeria had comparable per capita incomes, and both were heavily dependent on oil revenues. Indonesia's per capita income is four times that of Nigeria. Nigeria's per capita income has actually fallen, from US$302.75 in 1973 to US$254 in 2002'[8] Indonesia per capita reached 3,877.74[9] while the GDP when adjusted by PPP is equivalent to 61% of the world average whereas the GDP per capita of Nigeria is equivalent to 19% of the world average. The GDP per capita is average 1648.26 USD from 1960 until 2016, reaching an all-time high of 2563.10 in 2014 and a record low of 1089.10 in 1968.[10].

Clearly, Nigeria's vast oil wealth has not translated into the promised riches nor is a source of blessing for a clear majority of the people; neither has it led to improvement in the standard of living of the majority as widely expected. With prevalence of the 'Dutch Disease' Nigerians cannot totally demonise oil. In a market economy, all commodities are subject to market forces and volatility. Still, market is a force recalcitrant to intrusion to its total control of the environment. Given oil mercurial nature, it brought with it some positive developments in health care, transport, education, communication but they all came at a price: the ill-health of the environment. Despite the deficiencies oil brought improvement in GDP.

Again, there is a drawback to oil vital role in the economy. The downside is oil has magnified the rift between rich and poor Nigerians. With the market induced slump, hard won improvement in GDP took a major blow from rhythmic 'boom' and 'bust' cycles of the oil market. And after every cycle the country plunged even deeper into a more vicious cycle of poverty.[11] Statistics bear out how the numbers are plummeting with rising incidence of poverty. Accordingly, the source of economic fortune is also undermining the environment modern wealth created. Though Nigeria's carbon driven economy succeeded in reinvigorating its fledging wheels, there is a caveat: it works at cross-purposes with nature hence other global initiatives targeted at reducing greenhouse gas emission.

Stepping back from this sequence, if Nigeria and the rest of the developing world are to embrace the high standard of living of mature economies with its negative entropy, emissions from these countries could negatively tip the GHG scale. Developing nations might stop at a level of about 2 kW of primary energy per person.[12] Basic human need are satisfied and consumption at this level. Nevertheless, it would be only a fraction of Europe or United States rates[13] such policy places a new barrier on least developed countries from harnessing their vast natural resources to bring improvements in the standard of living of their people. In doing so, a road most matured economies travelled have now become a major source of decarbonisation; an action some argue tantamount to reinforcement of underdevelopment and concretization of dependency.

In examining this carbon catch-22 impasse, Klaus S. Lackner is of the view that low income countries have not caught the high consumption pathology of developed economies. Even so, carbon dioxide content of the global atmosphere is rising at the rate of 2 ppm per year. What this view objectifies is given current wave of development in developing economies, their consumption will increase hence the uptake of carbon is here with us.[14] With adverse footprint of the global economy, the spectre of climate change which is the outcome of our saturation of the ozone layer with GHG now is a global reality. Lessening the carbon intensity of the global economy requires reducing carbon uptake to an acceptable threshold considered safe for human existence of between 400-450 ppm.[15].

Unfortunately, Nigeria has caught the bug of emitting high quantities of carbon dioxide since it is afflicted with 'the original sin' of having vast oil endowment extracted to improve the living standards of the people. It is on this premise that some critics wholesomely blame fossil fuels extractions for Nigeria's legion of woes. The vast fossil fuel deposits have no blame for the pollution intensity of the economy. If we are to lay out any blame, we should place the blame squarely at the doorstep of oil industry operatives and their growing army of rent-seekers who have hijacked extraction and consumption processes.

Common Solution for Climate Change and COVID-19

Investments decisions of today are vital to how we shape the future hence are material to how decarbonization of the biosphere unfurls. How can the world halve its carbon uptake using the opportunity COVID-19 avails? Can we continue on the carbon-led part hopeful other pandemics and shocks will not erode the gains in human progress? Though the global recovery packages will cost billions of dollars but will the country use the stimulus package to stimulate low carbon development? What should Nigeria do to seize this opportunity or will she wait for another climate shock to place reprioritization of nature on the front of national conversation?

Two major risks are the defining challenge Nigeria will grapple with in the foreseeable future. Today's risks are climate change and COVID-19. Collectively, they are defining how we relate with one another and with nature. These risks though mutually exclusive are complexly interconnected. Taken together, they are not just existential threats requiring urgent attention but are placing additional stress on the revenue generation capacity of the country. Another common ground is the need to build a strong economy to limit devastations of these risks. For Coronavirus pandemic, improvement in public health services are necessary adjunct of life whereas in the climate change tuff, low carbon economy answers the constraint of reducing GHG emission to the acceptable 1.5 °C threshold. Realization of these diverse goals requires extensive capital outlay and minimization of carbon upload.

Inexpensive energy is necessary for economic advancement for recovery from the staggering economic effects of Covid-19. Hence Nigeria's post-COVID-19 recovery stimulus must strive to reposition society towards the level of economic activities existing in pre-COVID-19 era. Arriving at this threshold would require high carbon uptake outstripping the level of GHG emission recorded before the pandemic. To put the economy back to prior state entails burning carbon eroding the narrow gains in carbon reduction recorded in the COVID-era. This requires trade-off between economic growth and energy transition. Going by the wave of stimulus fleshed in the country, the trend is building enduring structures sufficient in avoiding eco-costly fossil fuel-led growth towards a low carbon and climate resilient transition. Emphasis is placed on growing the economy in line with NDC to minimize waste while also making the economy climate resilient.

However, prioritizing climate policy will harm the ability of most people to improve their living conditions, particularly after the terrible economic shock caused by the lockdown. The contradiction inherent in prioritizing climate policy is inevitable. Again, if the country has less wealth due to reduction in burning carbon then they will have fewer resources for environmental protection neither can she confront the public health scourge frontally. The issue here is whether those clamouring for greater economic growth hate the planet. The real question here if environmentalists hate humanity and the planet too. Therein lies the rub. Ideally, a low-carbon economy provides an opportunity to emerge from the COVID-19 pandemic while addressing the second defining crisis of our era - global climate change. Again, burning less carbon would undermine efforts seeing that the economy will not produce enough carbon to generate energy to a level commensurate with requirements of growth.

If the world must reduce greenhouse gas emissions by 7% each year and 45% by the end of this decade, then COVID-19 presents an opportunity to get their acts right. Additional investment is required for the stimulus package to kick start growth but care must be taken it does not promote unbridled growth from carbon intense infrastructure. Meeting the NDC and economic recovery requirements simultaneously is double imperative. Notwithstanding, it must be such that aligns priorities towards low carbon technology. Investment in six areas – renewable energy, energy efficiency, low-emissions transport, sustainable land-use, food loss and waste, and improved urban development is the key to a prosperous and eco-efficient future. With this, fossil fuel dependent economies like Nigeria are coming to grips with the dilemma of either confronting their carbon addiction head-long or face the menace of COVID-19 using the most tested path to growth. The crux of the matter is whether Nigeria can jettison the country's recovery in pursuit of environmental quality.

An avid apostle of using COVID-19 recovery to kill two birds with one stone is Christiana Figueres. In her opinion the most consequential question looming over us today is not whether we can address the Covid-19 crisis and climate change at the same time, but rather whether we can afford not to do so. This argument sets the tune of post-COVID-19 recovery stimulus which most mature economies keyed into. The reason is obvious. They have the prosaic goal of using the large fund injected into the economy to propel growth towards new development that serves dual mandate of decarbonization as well as economic recovery.

This owes to COVID-19 pandemic colliding with climate change emergency. Since they all require urgent attention for minimization of these mutually exclusive global scourges, it is therefore inevitable that plans to curb the pandemic must also have climate change mitigation as an achievable end. Given this, there is the imperative to integrate the solutions to both crises into a coherent response. In the aftermath of public health challenge, an inclusive recovery programmes must propel the global economy towards sustainable growth and increased resilience. Although the world in post-COVID era is on the path of rapidly growing fossil fuels infrastructure to recover from coronavirus care should be taken not to turn it into another road to ecological perdition.

Fatih Birol of the International Energy Agency makes a strong case for clean energy transition in the post-COVID era stating that "we should not allow today's crisis to compromise the clean energy transition". Governments should scrap $400bn in fossil fuel subsidies and back energy efficiency, as well as clean energy and infrastructure.[18] In similar vein, Stéphane Hallegatte of the World Bank highlights potential investments deemphasizing carbon-led growth such as restoration of degraded lands, sanitation and sustainable transport infrastructure.[19] According to Hallegatte, governments can use this to create millions of jobs in the short term, spur innovation, support economic diversification, and cut both carbon and air pollution and improving public health.

Notes

1. WHO (1999) *Influenza Pandemic Plan. The Role of WHO and Guidelines for National and Regional Planning*, WHO, Geneva, 66 pp.
2. McKinsey & Co (2020) Virus Puts 150 Million Africa Jobs at Risk, McKinsey Says. Accessed April 18, 2020 from https://www.bloomberg.com/news/articles/2020-04-16/virus-impact-may-affect-a-third-of-africa-s-jobs-mckinsey-says
3. Okoh, A. I. S. (2019) End of Fossil Fuel Economy: Weathering the Storm of Transition to Post -Fossil Society in Nigeria. World Review of Science, Technology and Sustainable Development. Vol. 15, (3) pp. 197–213. https://doi.org/10.1504/WRSTSD.2019.102110. According to Okoh, exit from all is now inevitable given that most economies are travelling on the road of renewable energy.
4. The Vice President Yemi Osinbanjo address to the 55th anniversary of the Oil Producers Trade Section of the Lagos Chamber of Commerce and Industry (LCCI) in Lagos said Nigeria must use its oil wealth to prepare for a future when the world no longer runs on fossil fuels.
5. Okoh, A. I. S (2017). *Pathways to a Green economy in post-Paris Agreement Africa.* Green Economics publishing House, Oxford. Available at: https://www.greeneconomicsinstitutetrust.org/books/pathways-green-economy-post-paris-africa/

6. The New Climate Economy: Seizing the Global Opportunity. Available at: https://2015.climateeconomy.report/
7. Joseph E. Stiglitz in 'Making Natural Resources into a Blessing rather than a Curse'.
8. https://www.ceicdata.com/gdp
9. https://tradingeconomics.com/norway.
10. Collier, P. (2008) Oil, Growth and Governance in Nigeria In P. Collier, C.C. Soludo, C. C. and C. Pattillo (eds.), *Economic Policy Options for a Prosperous Nigeria* New York: Palgrave Macmillan
11. Ibid.
12. In KLAUS S. LACKNER (2010) Comparative Impacts of Fossil Fuels and Alternative Energy Sources reference is made to the publication of V. Smil (2006) in OECD Global Science Forum: Conference on Scientific Challenges for Energy Research, Paris, 2006. and. B. McKibben, New York Times Magazine, July 23, 1995.
13. V. Smil, Annu. Rev. Energy Environ., 2000, 25(1), 21–51.
14. Klaus S. Lackner (2010)
15. The issue of an acceptable temperature was hitherto an unresolved issue but since the IPCC (2007) the scientific community have agreed the global temperature of 2 oC is the acceptable threshold by this it means essentially that emissions should be the 400 to 450 ppm. See IPCC- Intergovernmental Panel on Climate Change (2014). *Fifth Assessment Synthesi (longer) Report*. Retrieved from IPCC:www.ipcc.ch/pdf/assessmentreport/ar5/.../SYR_AR5_LONGERREPORT.Pdf
16. The International Energy Agency predicts that the Covid-19 led economic recession will reduce greenhouse gas emissions this year by 8%. https://www.iea.org/reports/global-energy-review-2020/global-energy-and-co2-emissions-in-2020.
17. Available at: https://www.epa.gov/sites/production/files/201612/documents/sc_co2_tsd_august_2016.pdf
18. Fatih Birol head of International Energy Agency https://www.theguardian.com/environment/march/2020/12/coronavirus-poses-threat-to global-activities-so-says-watchdog
19. Hallegatte of the World Bank Available at https://preventionweb.net/news/view/71103

Chapter 3
Political Economy of Fossil Fuel Exit

'I call petroleum the devil's excrement. It brings trouble…waste, corruption, consumption, our public services fall apart, and debt - a debt we shall have for years…'
Juan Perez Pablo Alfonso, the Co-Founder of OPEC (the Economist, May 22, 2003).

A Time to Change the Game

We are living in an era of one of the greatest mass extinction spasms with climate change as an interdependent phenomenon which is accelerating this geological contraction. We are living in an era of global interdependency with global warming an interdependent phenomenon humanity created and thus must contend with it. The culprit according to experts is the warming of the planet and humans are its major causation.[1] Our world is warming at an alarming rate that one may agree with Joseph Stiglitz who said the world is acting out the doomsday's script with key global power actors as the script writers in the climate endgame.[2] This endgame is motivated by the all-encompassing integration of the global economies drawing humankind dangerously towards inflicting irreversible damages to the global common resources. The velocity of environmental change recorded is taking an apocalyptic dimension.[3] Overwhelming body of scientific evidences clearly indicates that this climate emergency or climate breakdown is a serious and urgent issue.

The Earth's climate is, indeed, rapidly changing mainly because of increases in emission of greenhouse gases caused by human economic activities. Most climate models show that a doubling of pre-industrial levels of greenhouse gases is very likely to commit the Earth to a rise of between 2–5 °C in global mean temperatures. This level of greenhouse gases will probably reach between 2030 and 2060.[4]

This is particularly so in most countries undergoing rapid industrialization. Rapid industrialization is leading us dangerously towards climate change or better still, a climate crisis. In the analysis of the Intergovernmental Panel on Climate Change (IPCC) Special Report on Climate and Land (2019), climate experts warned that without any actions, the world is already on a pathway towards crossing the Climate

A. S. Okoh, *Oil Mortality in Post-Fossil Fuel Era Nigeria*,
https://doi.org/10.1007/978-3-030-60785-2_3

Redline as the impacts of climate change are already outpacing climate actions. And as the impact of climate change manifests daily, it is changing the what, when and how politicians and other players in the energy, agriculture and other high cárbon intensive industries act to mitigate the irreversible damage.[5]

No doubt, climate change is changing socioeconomic and political processes globally. Manifestations of global change have immersed sub-Saharan Africa (SSA) in sustained ecological crisis which is leading to market failure.[6] These market related crises have increasingly necessitated constant reconfiguration of the socioeconomic landscape; redefining existing relationship between states and non-state actors in Africa on one hand, and between Africa and the developed economies on the other hand.

The impact of climate change in a country like Nigeria is not limit to economic aspect but is a two-way interactive process of modifying and redefining the environment of policy and climate alike. A major game changer accelerating GHG build-up in Nigeria today is climate change. Climate change has devastated most communities rendering many people homeless. Efforts to curb this menace at international level are material to national policies formulated to address this scourge with dissimilar interests pulling international commitments towards diverse ends.

Premised on this, climate change negotiations are keenly contested affairs. The contests between different self-interested parties made HE Teburoro Tito, President of Kiribati to see the Conference of Parties (COP) to Kyoto Protocol as a messy compromise from an uncompromising interest group with diametrically opposed views all pulling negotiations towards their goals. As HE Teburoro accurately puts it, the world community are behaving like 'greedy piglets fighting over their share of milk from their ill mother. Instead of co-operating on how they could help their mother, they were all carried away fighting for their share, and making their mother even more ill, until all of a sudden the found that they have lost their mother and there were no more milk to live on'.[7]

Though there are remarkable improvements in international negotiations since the first UNFCCC (COP1), the world remains fundamentally unchanged from the 'musical chairs' of blame game on who is the major contributor to climate change and which hemisphere contributed more to upload of carbon. Unfortunately, the blame game has not really changed. What has changed is the intensity of the blame game with the real substance largely unchanged. Now the NDCs under the Paris Agreement squarely lay the responsibility of servicing the climate debt on every country whether developed or developing country. To date, the medium through which developing economies strive to elicit restitution is through the backdoor of climate finance which is contrary to past practice of requesting for reparation for climate injustice.

Effectively, what developing countries failed to secure through the front door in the form of restitution is paid by developed countries in the guise of climate change finance. Hence a new name has been given to the request for climate restitution to close the global adaptation gap. Now more than ever, UNFCCC negotiations are becoming far more complex with national interests in least developed countries merged into one. Even though a strong undercurrent of disagreement has emerged, the borderline between countries of the global North and those of the global South is narrowing lately with the Paris Agreement.

Unarguably, the world is moving in the right direction while in some aspects we are also in motion without movement. Indeed, the world is moving in the right direction, but on varied trajectories where the weakest species are at the receiving end of the new wave of unilateralism based on populist and nationalist (right wing) governments coming to the centre stage at global level. We are witnessing emergence of dissimilar forces pulling the world towards diverse directions with some of these countries inclined towards niche construction. According to climate experts, the global geopolitics, the economics of climate change and control of global power continue to be clogs in the wheel of climate change negotiations delaying urgent actions needed to tackle climate change while the withdrawal of the United State of America from the Paris Agreement only serves as salt on the already inflicted injuries.

Though all species of the animal kingdom are engaged in constructing niches, humans (most especially developed economies) have excelled in construction of niches than all other species put together. Human niche construction is highly reliant on energy which is causing more greenhouse gas uptake. The urge to accelerate niche construction is high in the US lately, for instance. Although niche construction is human embedded in our DNA to build new frontiers in extreme conditions but using social construction of one nation to undermine another negates WTO principles.

In such a clime, the "struggle for existence", "Competition", "growth" and "optimisation" is the centrepiece of man's existence. Selfish pursuits and corporate interests are machines which drive innovation and not planetary well-being as we are made to believe and it is not in the interest of nature or the threats of climate extreme events. In this perennial competition, fit species (powerful corporations) exploit niches (markets) and multiply their survival rate (returns margins) while weaker (less efficient) one's going extinct (bankrupt).[8]

In the last three decades, mankind's plundering of natural resources immensely increased making scientists to describe this era as the Anthropocene, or Homogenocene. (The Anthropocene is an era of positive or negative technology driven innovations). In this era of high mass unsustainable production and consumption, various technologies have emerged to reduce harm posed the planet. Humanity, no doubt, is never in a short supply of solutions hatched to bail us from existential challenges. The merging and clashing of these technologies at different era found a way of re-emerging in a more complex way seeing that new avenues of coexisting with the old would manifest until after a time when the old technology either fizzles out or market forces made such technology unprofitable.

Three interrelated global instruments drive radical changes in the world lately: the Paris Agreement, Sustainable Development Goals (SDGs) and the Sendai Framework for Disaster Risk Reduction. The overarching goal of SDGs #7 is to generate energy for all and end all incidences of energy poverty in 2030. Attainment of energy security is therefore the fulcrum of this agreement owing to the crossing-cutting impacts on all sectors of the economy. If the plan to generate energy for all by 2030 should proceed as planned, it might require additional infrastructures invariably requiring increases in fossil fuel energy generation while using natural gas as a 'bridge technology'. There is insignificant robust energy system to revise power generation deficit given that the existing infrastructures are not sufficient to bring about desired improvement in the energy sector especially in most of the developing countries. Building energy infrastructure of the next two decades is, therefore, critical to a sustainable energy future.

The impetus for this radical transformation is partly attributable climate change. Another conundrum is economic expediency. Within scientific and political circles climate change is a global reality adjudged the defining challenge of modern age. For this reason, there are concerns it will undermine development in Africa even though the continent contributed only 3% of the global emissions.[9] Emission from carbon dioxide is set to ratchet up the impact even further thus increasing food, water and energy poverty. In doing so, it is adding significant uncertainties to production and consumption of natural resources. Therefore, the imperative to set the bar on ambition to reduce fossil fuel consumption since climate-related risk is exacerbate by means of socioeconomic and political factors. Some might argue this invoked a new age of demagoguery wherein eco-efficient solutions hijacked to serve other ends. Reduced to its essence, the global decarbonisation agenda is another opportunity for rent seekers to exploit nature. In this rentier paradise, trend towards using renewable technologies will offer staggering returns to whosoever grabs the castle first (new renewable niche market) with workers who produce the wealth of the nation caught holding wrong end of the stick.

The global solution to correct this mutually exclusive phenomenon is to agree on a series of conventions and protocols such as: Kyoto Protocol and now the Paris Agreement. This is the result of another equally significant landmark intervention-the United Nations Framework Convention on Climate Change (UNFCCC). From the first Conference of Parties COP1 in 1992 to COP25 in December, 2019 in Madrid diverse far reaching commitments on the decarbonisation have been discussed. The primary goal of this commitment is to initiate policies enabling different countries embark on a low carbon emission pathway; in this way reduce the average global temperature to 1.5 °C and well below 2 °C (degrees Celsius) threshold considered safe for human existence.[10] Even then, there are still some problems which marred the outcome of the recent post Paris COPs.

A major talking point at COP24 was the wordings of the IPCC Special Report on 1.5 °C global warming most especially the use of the words 'welcoming' which some fossil fuel exporting nations think should change to 'noting' when describing the need for decarbonization. With the compromise on the exact wordings of COP24, the views of fossil-fuel lobbyists seem to have held sway. A major gain of the IPCC report reveals that global temperature is moving towards a catastrophic rise in the coming year. Thus, the world requires immediate and far-reaching changes in upload of carbon into the atmosphere if we are to avoid near-term climate catastrophe.

A key measure adopted in the report to avert dangerous climate change is a 45% reduction in the GHG emissions by 2030. The world leaders' attempt to reduce human interference with the atmosphere in form of rebalancing biogeophysical properties of the earth system is a complicated matter. Not only is carbon emissions and economic growth closely interwoven and as such inseparable, but they are also inversely related with the political fortunes of governments in some weak states. Surely, it is closely related to the political fortune of the party in power hinged on the fact that human activities that engender emission of CO_2 also lead to major developments in most countries. A glance at historical trend shows how combustion of fossil fuels (coal, natural gas, and oil) for energy and transportation in Northern countries fuelled rapid transformation achieved in the eighteenth century.

Historically, natural resources exhaustion and agro-ecological deterioration take place concurrently linked to population explosion just as they are equally associated with the Industrial Revolution. Then, the global interest was not focus on curtailing the negative effect of our carbon output rather was slant towards promoting growth into the infinite. With the establishment of linkages between carbon-based growth and population explosion, it is increasingly clear decarbonisation process can either exacerbate or alleviate poverty level experienced in countries. It is, therefore, tautological to say resources are becoming scarce with population explosion. To a certain extent depletion is rising with innovations in technology, technological pace is not catching up with the exponential population explosion. Technology is bridge between closing the energy and resource gap. Consequently, rapid transition from our present anthropocentric vision towards a more biocentric/eco-centric view is taking place in the process moving the world away from carbon-emitting energy infrastructure towards a carbon–neutral infrastructure.

The result of the ongoing technological shift is the gradual loss of oil's share of the global energy mix. For instance, in the past fifteen (15) years in a row, oil's share of the global economy, gradually declining with cost competitiveness of renewables such as solar and wind remarkably growing.[11] Another factor propelling the global economy away from fossil fuels is the IPCC Fifth Assessment Report. The IPCC Report came up with the important finding that reducing the global temperature through total elimination of fossil fuels is necessary (by 2050) if the world must avoid further degenerations.[12] Given this and other related factors, drive to find other alternative sources of energy gained further momentum. The Paris Agreement is the catalyse of the renewable energy revolution in doing so is making the fossil fuels sector less attractive to investors. Accordingly, an increasing pace of divestment from oil is happening with over 1000 corporate entities totalling $6 trillion withdrawn from fossil fuels infrastructure towards renewables.

With this investment, massive efficiency revolution in solar and wind energy sources is underway. But here is the caveat: Figures emerging from varied scientific findings on this transition must be with caution. Although these figures depict actual rate of divestment, it is not totally over for the fossil fuel sector. One thing we can say with a high degree of confidence is the world heads for more severe times in the years to come. This is for both oil producing and non-producing nations with the scale skewed against the former. For the global economy to make any serious dent on the carbon upload, then efficient utilization of resources must march economic growth in the transition to a low carbon future. For this reason, energy access and energy transition must receive the same treatment. Sorry to say, the present scheme of events tilts towards building fossil fuel energy infrastructures with some level of efficiency thrown in.

It is obvious the gap between energy access and energy transition in developing countries is growing at an alarming rate despite remarkable growth in deployment of renewable alternative energy. Huge gaps exist between what is produce, those with access to energy and how these countries transition to a low carbon future. Clearly, plethora of problems bedevils the drive to close the energy access and energy transition gap impairing realisation of dual mandate of the global community. Even though price signals that are economically efficient to allow timely adoption of low-carbon technologies is ongoing; it will not improve the standard of living nor will it minimize the energy access shortfall.

Nigeria has an acute energy access crisis (in the energy access model, the International Energy Agency (IEA) defines initial electricity access as 250 kWh per year for rural households and 500 kWh for urban households with the projection of this base level increasing to 750 kWh per person in 2030). Alas, the access level in Africa is about two orders of magnitude less than the average American consumes on a yearly basis. In Nigeria, the gap between energy access and energy transition is widening while population growth outstrips the power generated. With the energy access low, resolving the dilemma of universal access to electricity (SDG#7) will task governments most especially in Nigeria where a plethora of constraints are prevalent.

What is unfolding in the global economy can be termed a clean energy revolution with accelerated deployment of energy efficient technologies. In this age, new power players within the energy sector are determining the direction and pace of the uptake of technology through their control of low carbon industries. These new key players' primary motive is to close the low carbon infrastructural deficits with fossil fuel power generation. Success story of the likes of Tesla in promoting take up of electric vehicles is a testimony to this. Nevertheless, removing fossil fuels from the global economy is not a walk in the park because eliminating fossil fuels from the transport sector could take quite some time. No easy answers are available for decarbonisation in a large economic sector wherein shipping, air travel and chemicals utilization of fossil fuels are high.

Pressures to completely remove fossil fuels from the global economy will grow, although action is not being taken quickly enough to avoid dangerous climate disruption', so says Donovan[11] and other climate experts. What this objectifies is we have not heard the last of the dying dinosaur. For the world to eliminate this dinosaur, then, all the leaking holes in the sinking ship of fossil fuel energy generation through large economic sector (chemicals, shipping and air travels) are eliminated. Besides, crude oil producing countries like Saudi Arabia, Russia, UAE, and Nigeria will not let go without stiff resistance as they are incline toward achieving the two divergent ends concurrently.

For instance, Poland who hosted the COP24 is also the coal capital of the world. Paradoxically, this same country which played host to the global community on eliminating fossil fuels from the global energy mix is also pushing for the survival of their main income earner- coal. Whereas other countries such as Nigeria is reforming their fossil fuels extraction through new petroleum policy on ending long-running systemic failure of state-run organization (NNPC). This effort instead of curbing unpleasant pollution is pushing the economy towards carbon intensive growth and consequential desecration of the ecosystem.[13] The end result of such rebranding in the oil sector will be an upsurge of pollution when all leakages in the smooth extraction of oil are all plugged or it could concretize fossil energy market since it will redefine the way the country consume energy thereby fundamentally altering the energy systems. Contrariwise, restructuring of the moribund oil industries can change the march of history in the deployment of energy efficient technologies. When cheap and smart renewables are readily available, efficient and stable then, we can provide energy when we want, how we want, and as close as possible to its original cost.

Economics of Natural Resources Production

To grasp the political underpinnings of fossil fuel and circumstances surrounding its exploitation, it is necessary one is conversant with some key green economic concepts. A general overview of related concepts such as pollution, common property, public good and externalities is a required to really grasp the economics of climate change, why concepts such as the polluter pay principle (PPP) have come to stay and how it dovetails into the global emission reduction initiative.

Pollution is a problem of scarcity in terms of waste disposal capacity. The challenge is of choice mostly of how to utilise scarce resources in relation to societal needs. Some have held that markets are helpful in determining how these scarce resources are utilised in the most rational manner. For eco-equilibrium, there must be equality of demand and supply of environmental quality. In a strict economic sense, pollution is the outcome of any loss to human well-being arising from physical environmental change. Pollution arising from environmental change may have short-term or long-term impacts upon the health of human beings.

In considering natural resources, pollution is degradations of the human environment.[14] In this understanding, other forms of degradation are exploitation of natural resources other than for cropland, exhaustion of non-renewable resources and mismanagement of renewable resources. Whilst for decision-makers pollution is all negative impact experienced by users of the environment because of degradation. Impacts of pollution have diverse categories differing from one environmental milieu to another. These are air, water, and land pollutions with all these having dissimilar ramifications on man. Besides the damages human beings may incur, pollution can heighten environmental destruction thereby encouraging obliteration of genetic information in plant and animal species.

Pollution of any kind is therefore an environmental and economic problem requiring us to make a choice and to resolve conflicts of interest. This is a situation only policy must address. Every type of pollution is the outcome of interest of one economic agent at the expense of another. Resolving conflicts of interest between economic agents is the end of most pollution ameliorative policies. However, political decisions allocate resources to reduce pollution intensive activities. The process of reducing pollution will either improve the environmental quality or reduce the value of some resources at the disposal of society. Paradoxically, however, environmental assets cannot produce all human requirements while achieving eco-equilibrium for the mere fact that producing one good will be at the expense of another. The problem of choice arises when using environmental assets to grow the economy or in ensuring eco-equilibrium through maintaining environmental quality.

Resolving this dilemma is the task of policy. In this, the policy process faces conflict between two ends: environmental quality and economic growth. And where there is conflict there must be a potential gainer and loser. In this case, the environment is the loser with man bearing the ultimate burden of all environmental losses. A burden to policy is the undesirable effects of industrial production which invariably affects the environmental quality. The goal of policy for these environmental complexities is how to restrict those production activities placing various demands on the ecosystem. To mitigate damages resulting from pollution, environmental quality is weight up against perceived gains. Efficient functioning of the ecosystem is very important because damages caused may affect assimilative capacity to handle waste disposal which, again, will stretch mother Earth's carrying capacity to the biophysical limits (The term assimilative capacity refers to the ability of natural system to accept certain pollutants and render them harmless).

Some environmental resources may have assimilative capacity enabling them to absorb wastes thereby rendering them harmless. Examples of these are the ability of water to absorb some waste heat without degradation, the decomposition of organic materials through natural processes and the dispersal of atmospheric emissions using air currents and wind patterns in levels that are harmless (i.e. they will have no measurable effect on directly consumed, environmental services or on production processes).

When output increases, it may accompany sustained increase in recycling and resource recovery. This is possible if there is an increase in the relative price of virgin raw materials, perhaps using severance tax, or a policy of charging the full social cost for waste disposal in an environmental medium. Additionally, if virgin materials are commercialised, increasing scarcity may induce a price increase, encouraging recycling. An increase in recycling and resource recovery will decrease the throughput of virgin materials; reduce waste flows and moderate stress on environmental resources in extractions. There are several policies targeted at challenges impairing regenerative growth potentials of the environment in Nigeria. Even then, there is a huge gap in their implementation. The gap in formulating policies with the wellbeing of natural resources is due to rising dependence on natural capital. Within business and political circles, there is general perception of the ecosystem as having a regenerative capacity hence incentive to limit degradation is insignificant. While others urge an increase in emissions would result in further damages.

Still, some do contend that if emissions decrease, then the damaged ecosystem will go back to previous levels. This reasoning is fundamentally flawed given the diverse life cycles of resources although this assumption might hold for some damages. In general, an element of irreversibility is attached to most ecosystems for we know an ecosystem once degraded is most times irreparable or takes very long time to heal (400 years for polythene-based pollutants). Maintaining the delicate balance of the Earth's ecosystem and biodiversity should be the primary goal of all policies fashioned in our environmentally hostile world.

Humankind is extensively dependent upon natural resources for various benefits with these ranging from fibre, air, clean water, healthy soil and carbon capture and much more. Most of these resources are global common properties deemed to be public goods in which market and price is not attached to them. Because there are no prices attached, they are rarely detectable through economic compass. The concept of public goods presupposes non-rivalry in consumption and non-excludability. Here, non-rivalry means consumption of a good or service of one party does not diminish the availability of the goods or services for others. Whereasnon-excludability, on the other hand, means for technical or cost reasons, it is not feasible to exclude consumers from using the goods and services once produced. When a consumer is not excluded from using such goods or services; there is no incentive to purchase them.

What follows then is the potential producer with no prospect for sale or profit will have no incentive to bear the cost of production and little or none of the goods or services will be produced.[15] Consumers would, in this situation, act as free-riders. A free-rider is a person who underpays relative to the benefits he derives hence it is a tragedy of the commons. If there are no incremental costs to supplying the goods or services to additional consumer (non-rivalry in consumption (no opportunity or user cost), the appropriate economic price for the good or service is zero since once produced; it would be impossible to exclude anyone from its use. It follows then that private market system will fail to produce an appropriate supply of public goods. An example of public goods is the air. If the air is clean up for one person in an urban area, it is automatically clean for all in the community. The benefits of the clean-up accrue to all in the community.

Environmental quality is, essentially, a public good. Because of the free-rider syndrome, we cannot rely on market system to provide efficient quantities of public goods. We must rely on some non-market institutions involving a conceptual action of one type or another. Here, a form of governmental regulation or policy will be needed to address the problem. At present, environmental degradation and biodiversity loss are on the rise despite efforts from governments, NGOs and businesses around the world. Increase in degradation is linked to failure of market-based instruments to produce socially acceptable results. Failure to limit degradation is because markets tend not to assign economic values to the largely public benefits of conservation of environmental capital while assigning a value to private goods and services the production of which may result in ecosystem damage. In this sense, market failure is an environmental externality because outcome does not pass through the market.

The concept of public goods is closely linked to externalities. An externality is an incidental not necessarily anticipated effect caused by one economic agent on another where the effects do not pass through the market. Market failure is anything resulting from the lack of market for public goods and services sometimes called public good failure. A good example is the absence of a market for species conservation or for most of the regulating and supporting services of the ecosystem. This also includes imperfections in market structures or processes around the market which caused inefficiency and distortions. For example, the rise of carbon trade globally has given vent to further pollution since it legitimises pollution.

However, pollution from one economic agent on another demonstrates that not all externalities are incidental. Economic agents can be producers who contribute to negative or positive externalities in the economy. The reason for this is obvious: environmentally related pollution is a subgroup of externalities and has mostly negative attribute. A good example of a positive externality is the relationship beekeepers and orchard owners enjoy in the pollination of fruit trees. Such relationship is a positive externality because in this instance, both the beekeeper and owner of a nearby orchard benefits from the activities of the bees. When the impact adversely affects another, it is a negative externality.

In the example above, the connection between the concept of public goods and externalities becomes more evident. Interdependence of the beekeepers and orchard owner is a case of privately produced public good. When an item is a public good, it is not possible to exclude users. In positive externality, the agent producing the externality cannot appropriate or charge for its beneficial effects. Whereas with a negative externality, the victim is an unwilling recipient of the effect and cannot himself directly exclude the effect. The producer, in this case, does not bear the cost but shifts the burden to the victim. An example of a negative externality is the relationship between an upstream factory and the downstream users of a water source. The upstream factory discharges toxic waste through an environmental medium such as water. But for those who rely on water sources at the downstream, it is a case of negative externality because the waste generated by the upstream party is posing serious harm to the health and economic well-being of the downstream party. Given this, the lack of excludability is central to both public goods and externality issues. If the waste discharged in the water encourages fishes to grow then to fishing industry it is a positive externality but a negative externality for the water purifying firm who use additional chemicals to remove the impurities.

The major challenge in climate change mitigation is how to address conflicts arising from externalising cost to others. For an appropriate pricing mechanism; cost must be internalised and the polluter made to bear the burden of their emissions. This is the basic idea behind the Polluter Pay Principle. Because there are lax environmental laws and weak institutions in Nigeria, internalizing cost is the critical problematic of environmental governance. Corporate dominance of policy based on economic rationality and the myth of boundless economic growth using technological deliverance has assumed the mantle of harbinger of environmental salvation. This is by no means an accidental task but is linked to ongoing reorganisation of the global liberal market system. Since markets are being reorganised globally, vital components of our planetary well-being must also be addressed. By extension, man is undergoing further refinement hence must take into consideration the ability of present stock of ecosystem services to provide for human wellbeing.

Inherent in most policies formulated with the environment in mind are deficiencies foisted through lax regulations on mitigating climate change. Mitigation and adaptation to the harmful impacts of this externality may be an avenue for either profit or loss. So, in resolving externality crisis there are winners and losers who directly reap from the absence of an appropriate regulatory framework. In Nigeria where there are weak institutions formulating policies for resources use and control, externality crisis is the daily reality of the people. As a place where weak policies reign, businesses have no incentive to control their polluting activities hence are most likely to take advantage of such lax environmental laws. Absence or inadequate laws may be an incentive to produce negative externality whereas government can through disincentives in the form of pollution tax formulate policies discouraging negatively produced public bad.

The gist of this as far as Nigeria is concern is that acute doses of externality crisis have emerged in country's economic space. Activities of oil producing firms in the Niger Delta region are classic examples of public good turned public bad. When viewed along gains government derive from the polluting activities of firms like Shell BP, Agip, Mobil and Total, it is a public good to government if other activities/policies produce benefits for the degraded communities. When the accidental or intentional economic activities of the firms produce pollution (externality) and is not remedy through the market, it is then a public bad, the outcome of a negative externality. Characterised in this way, the burden of polluting activities of Shell BP when borne solely by the host community is a potentially conflictive situation. A case in point is the current conflict over oil spills in the Ogoniland involving Shell BP. This is a direct consequence of a negative externality where the market failed to remedy.

To stem such externality crisis, government should play the vital role of engendering an attitude of compromise critical for evaluating pollution and controlling adaptation to change. A strong distributive and allocative justice is needed such that community social responsibility of the firms involved in pollution intensive extraction mitigates while at same time creates an environment suitable for adaptation to changes fostered on the people. Efficacy of policy on adaptation to pollution is most times proportional to the regulating framework and institutions bestowed with the responsibility. Policy failure may arise when incentives encouraging harmful actions are the goal of both public and private policies. Tax incentives and subsidies can spur environmentally harmful actions from a business entity thus making the market work as an agent of destruction, even in places where stock of natural capital can offer a sustainable pathway to economic development.

So, externality is a hard-economic truth, but whether market-led instruments are the most viable option for redeeming all market imperfections or finding which option is the best constitutes a major source of debate at academic circles. Surmise that externality will either create wealth for a select few or ingrain poverty for most of the people. In avoiding this climate injustice, proper allocation of economic burden is crucial for achieving a low-carbon growth. Often, options available to governments may be far narrower than is visible to the ordinary man. No quick fixes exist given the involvement of diverse interest groups. Where varied interests pull at an opposing end, it is a recipe for crisis and where there is conflict; one may agree it is fundamentally an argument over action. Such action could be in the form of disagreement over the allocation of resources. Where externality crisis is prevalent, then, unending climate variability will manifest daily. In such a clime, the questions posed by Dessler and

Parson, are critical for unravelling the impasse: how shall we respond to climate change? Do the risks it poses call for action, and if so, how much effort—and money—shall we spend, and on what type of action?

In resolving this dilemma some have argued for internalisation of externalities when social and environmental values are the central goal of policy. Conversely, others urge for the market be unfetter to determine what is right for both human and non-human nature. Thus, market prices must reflect real social and environmental costs and benefits, so that the polluter bears the cost of pollution. Others hold contrary view claiming tax regimes and regulatory frameworks should be used to 'tilt the playing field', making 'good' things cheap and 'bad' things very expensive.

In the Nigerian externality impasse, the good thing and bad thing are all from the same source so how do you categorise an item 'bad' or 'good'? When there is allocative and distributive crisis the most appropriate option is the greening of the economy using stock and sink concept to point us towards a most rationale future. We shall presently examine these competing concepts.

Generally, the environment plays a crucial role of 'stock. As a stock concept, the environment can be measured through indices of air and water quality, ore deposits, genetic diversity and so forth. It represents raw materials flowing into production and consumption processes directly. These resources can be further differentiated into renewable and non-renewable resources. On the other hand, the environment also acts as a sink. In the sink perception, it is argued that the environment performs the vital function of absorbing carbon dioxide emitted via pollution while through decomposition it can also reproduce environmental capital. Nevertheless, this conceptualization of fossil fuels in stock mode is not complete if one does not look at its utilization as feedstock. It is a known fact that fossil fuels provide services other than energy. It is equally well documented that fossil fuel acts as feedstock in many industries with different bye-products. These bye-products comprises of: pesticides, Fertilizers, plastics, artificial leathers, pharmaceuticals, tires, lubricants, textiles, waxes and asphalt and other things the world relies on come from petroleum, natural gas or coal.

As of late, we have graphic details of how advanced horizontal drilling technology are now used to exploit oil reserves in newer level never contemplated. At a rapid rate the world is discovering new oil reserves in the deep sea, arctic and other regions far removed from human habitation causing the stock of oil to increase. Even then, it cannot build the requisite sink function of the ecosystem instead is depleting the stock. These days, exploitation of previously unreachable areas of oil reserves has grown in leaps and bounds. Consequently, reserves of shale gas are on the rise with new hydraulic fracking technology used in oil extraction. With high intake of water, sand and chemicals, fracking is now increasing the level of environmental pollution and is adversely rebalancing atmospheric GHG.

Given the increases in gas exploration due to improvements in hydraulic fracking, crude oil is becoming cheap globally with the fall in prices reaching an all-time low of below $25 recorded in 2016. The crash recorded within the oil market is a welcomed development to net importers of crude oil and is equally an ecological negative due to its proclivity of increasing fossil fuel consumption. Unfortunately, it is a source of economic stress in oil producing states. The good news for oil producing countries is that prices began to appreciate. In the global market, price rose daily in the region of $80 in second quarter of 2018. Even then, there was some level of uncertainty. True to oil's price volatility, the price at the international level is swinging dangerously today. Towards the tail end of the last quarter of 2018 it was in the region of $50 with this change in line with market forces.

Yet, again, oil price slumped to $70 in April, 2019. This is not music to the ears of oil importing states' ears rather is another bad news for these countries who have hedged their bets on a stable oil decline in prices by 2019. An increase in demand from emerging economies is a major factor driving the rise recorded with an increase of nearly 31% in supply over the past decade recorded. This rise in price of oil is largely an outcrop of the volatility of the Arabian Gulf which is pushing prices up together with cuts from the major OPEC exporter: Saudi Arabia. All these changes in oil price were minuscular when compared to COVID-19 and Russia and Saudi Arabia price wars of 2020 fostered crash of the global energy economy. Coupled with COVID-19 lockdown the price of oil in 2020 slumped to an all time low of $15 in April before picking up to $45 threshold by July 2020.

Contrary to the expectation of oil stock depleting, new technology is pointing humanity towards new revolutionary pathways for discovering new deposits. According to a recent study from a fossil fuel sponsored ivory tower, it will take more than 1.76 billion years for present stock of natural resources to be exhausted. Known reserves of most industrial non-renewable resources are growing and the resources are estimate to be higher than they were 44 years ago. Known oil reserves in the 1970s stood at around 600 billion barrels with the world consuming about 760 billion barrels in the years before the end of the century. Current global reserve is in the region of 1.2 trillion barrels. Does this then mean that there is no crisis or looming scarcity? Should this be taken as a temporary delay in the race to scarcity crisis? Or is it an affirmation of the market's faith in the economy finding substitutes for depleted resources?

Undeniably, these issues are manifestations of larger ecological complexities facing most developing economies today. Examples of these crises include: depletion of the ozone layer, pollution of rivers and lakes and collapse of some fish stocks, etc. The question then arises if these are symptomatic of market failures or is an inevitable component of growth hence there is no crisis ahead? But human history informs us of the inevitability of crises such as those recorded in the environmental and health arenas (Ala COVID-19). Overcoming these environmental crises require corrective measures through good will and cooperation which are not readily available. To think there are no crises is to misinterpret history of human progress.

Said, oil has now become the new elixir but it is equally a doomed resource cursed to expose Nigerian fossil-led economy to the geostrategic machinations of developed and emerging economies. The reason is simple: In most Nigerian communities the rural poor are dominant population. And of these, 80% are engaged in agriculture-related sectors. Ditching oil will hurt these communities in the short-run but a transition to green economy would provide resources for sustainability of their livelihoods. The biomass sector in the long-run will be the dominant sector since these societies desperately need a 'Trojan horse' to provide them a safe passage out of the risky fossil-led economy and the COVID-19 pandemic. For this reason, preservation of ecological assets is mutually beneficial to the economy and environment hence Nigeria's transition would be based on social imperatives and not economic importance.

Emergence of Green Growth in Nigeria

In the aftermath of the Rio +20 summit ideas of greening the global economy to bring about the world to a net zero GHG emission trajectories flourished. Since then, many governments imbibed the green economy tenets without the requisite green trimmings in their Political, economic and environmental activities. Most of these states (Nigeria inclusive) attempted to operationalize the green economy concept within the ambit of sustainable development without adequately exorcising the ghost of the concept they are trying to supplant. In line with the Rio +20 outcomes, African leaders crafted a continental framework for reaching the global decarbonisation target. Several of the declarations and resolutions of these leaders strived to uphold green economy as an important toolkit for downscaling GHG emission.

Unarguably, localization of green economy principles in the strict sense of the word is, however, still work in progress. But in Nigerian context, no clear-cut distinction exists between the policy and practice of sustainable development and low or zero carbon economy. Distilled further, the concept of low carbon economy has not really gained currency within political circles neither is there any visible presence within academic circles.

Though green economy has some measures of currency within the political circle, a low carbon economy has not really entered the lexicon of national environment change governance neither has the circular economy concept. Green economy in its present form are activities pursuant of the principles of sustainable development or acts adjudged transitional towards downscaling GHG to a zero level or net negative emission. Expanding this argument, what is a low carbon transition exists in isolated sectoral policies housed under different ministries and organs of government which strived to green some sectors. Even then, formalization and integration into the national development plan in the strict sense is certainly not on board.

A case in point is the GGWSSI. The GGWSSI has institutional backing because it is with the admirable goal of greening a large swathes of land in Northern fringes of the country prone to desert encroachment. Undoubtedly, GGWSSI activities would reduce the level of GHG with the co-benefit of it promoting sink and stock functions and improving the biodiversity. Transition to a net negative emission trajectory promoted at Paris strives for the stringent target of 1.5 °C temperature threshold. Impressive as it may sound, government is cognizant of its effects on welfarist programme of transforming the economy by means of provision of efficient and adequate power to drive economic development.

Suffice to know that government's position on decarbonisation has not really changed since its submission to COP18. In the National Report for COP18/CMP 8, government averred that 'since meeting the 2 °C imperative could lead to a lower global demand for fossil fuel and a possible drop in the price of crude oil, Nigeria is cognizant of the potential negative impact on its socioeconomic goals of this response measure, despite the high degree of uncertainties'.[59] From this, one can infer the lack of political will to promote in a solid way the decarbonisation target set by the international community. What has emerged within government circles is one of ambivalence on decarbonisation with a non-committal outlook which hinges on economic exigency and political expediency and not environmental sustainability.

To demonstrate Nigerian government readiness and capacity to effectively address the likely impact of climate change, it crafted the National Adaptation Strategy and Plan of Action (NASPA) policy on climate change. The compilation of NASPA commenced from 2008 and was completed in 2012. The central plank of NASPA policy is not just to promote climate change mitigation and adaptation at the local level but, to act as a bridge between current disaster risk management efforts and reducing vulnerabilities to climate change at all levels. NASPA is Nigeria's adaptation blueprint vested with the principal responsibility to drive and coordinate as well as integrate all governmental efforts on climate change.

The ultimate goal of NASPA is to tackle the menace of climate change headlong. It requires the country, not only to embark on deep emission cuts but also to reactively and proactively implement coping strategies for the people. From its blueprint, NASPA policy seeks to minimize risks, improve local and national adaptive capacity and resilience, leverage new opportunities, and facilitate collaboration with the global community. NASPA aims at strengthening existing energy infrastructure by developing and diversifying secure energy backup systems to ensure that both civil society and security forces have access to emergency energy supply.

Accordingly, NASPA blueprint is the engine which drives commitments on adaptation to the adverse effects of climate change in Nigeria. However, the National Climate Change Commission Bill that should establish a National Climate Change Commission was passed in 2017 but is still awaiting the President's ratification. This lack of legislative framework seems to be an impediment to the implementation of adaptation response in the country. It seems that inadequate institutional support and inappropriate policies act as constraints to sustainable adaptation.

Nigeria also submitted her NDC in 2015. The NDC seeks to make an unconditional contribution of 20%, which is below business-as-usual (BAU), consistent with current development trends and government policy priorities. But will make a 45% conditional reduction with additional adaptation fund (FGN INDC, 2015).In line with the NDC, Nigeria developed and finalized the Sectoral Action Plan (SAP) for implementation of the NDC in key priority sectors of energy, oil and gas, agriculture and land use, power, and transport. The country also submitted the First, Second and Third National Communication required of all parties to UNFCCC. In line with the agreement the country also formulated the National Climate Change Action Plans (NCCAPs).

To recap, if acts intended at greening can be considered as a green economy, then, a form of low carbon economy is taking shape with some form of circular economy thrown into the bargain. But when juxtaposed against the prevalence of unsustainable gas flaring then the 'brown economy' is the dominant economic approach. The culprit, as already noted, is crude oil- the 'cash cow' of Nigeria's monolithic economy. It may be noted here, too, government has taken some bold steps in leapfrogging to a greener future using eco-efficient technologies but the speed and urgency attached to this uptake has not matched changes in the physical environment.

This point is worth emphasizing. Beset with manifold climate risks, national policies to stave off climate change are being formulated but they are not addressing the challenges as they ought to. Activities pursuant of the goal of decarbonisation though exist in blueprints/documents but are not fully ingrained into governance structures or concretely implemented to reduce the uptake of GHG. Inasmuch as the upload of dangerous gases continues unhampered then, climate change will also stand as sentinel at our doors. It is fact of marginal importance that policies on pruning down our carbon upload are formulated when they are not sufficiently operationalized to ensure the national decarbonization goals are met.

Energy, the Environment and Global Warming

Hydrocarbon extractions in the Southern parts of the country together with biomass consumption in the North are to blame for the negative carbon footprint.[16] The net effect of this is that climate change related risks are now commonplace. The International Food Policy Research Institute (IFPRI) estimates that by 2050 the combined effects of increasing temperatures, declining rainfall, floods and droughts could result in average rice, wheat, and maize yields declining up to 14%, 22%, and 5%, respectively. Whilst food availability in sub-Saharan Africa will average 500 cal less per person signifying a 21% decline. In the absence of adaptation measures, climate change could result in a loss of between 2 and 11% of Nigeria's GDP by 2020, rising to between 6 and 30% by the year 2050. This loss is equivalent to between N15 trillion (US$100 billion) and N69 trillion (US$460 billion).[17]

In addition to natural factors, man-made interference with the biogeophysical properties of the planet is to blame. Long term records shows how over the past 105 years, the average amount of rainfall per year dropped by 81 mm. So, emission of GHG and deforestation are the foremost anthropogenic causes of the phenomenon in the country. One key aspect of GHG emission is the carbon footprint noted to have contributed significant quantities of GHG to the atmosphere. For instance, Nigeria's gas flaring emits at least 400 tonnes of gases annually. Government has the potential to prevent carbon emission for as much as 3.7 billion tonnes over the coming 25 years. A World Bank study finds that even if just a fraction of this potential could be turned into assets tradable in the carbon markets of the future, revenue-generating potential would be significant.

Nigeria is well-endowed with a variety of energy resources. Oil production is the bedrock of growth with agriculture contributing 42% of GDP (Table 3.1). Putting an end to GHG emissions from biomass and fossil fuel energy is with severe consequences for the economic health of the nation.

But there is a further problem: Nigerian economy is also susceptible to external shocks. It took a serious bashing when crude oil price crashed thus unleashing further pains unto the people as a result is undermining the gains recorded in the last ten years (see Table 3.1). Furthermore, Nigeria's remarkable growth achieved when the national accounts statistics took place in 2014 indicates that the country was on track to achieve between 7 and 9% growth.[18] Unfortunately, the gains achieved at macroeconomic level have, however, been eroded through insecurity, rising unemployment, failing infrastructure, dwindling revenue from the oil and gas sector and climate extreme events. More significantly, two-third of the population who depend on agriculture is vulnerable to climate variability (Map. 3.1).

What is particularly disturbing is the rising incidence of climate change which is a direct threat to government's Agricultural Transformation Agenda (ATA), Vision 20:2020 and ERGP. Other implications of climate change on the economy include:

Table 3.1 Selected microeconomics indicators 2000–2009

Economic indicators	1990	2000	2001	2002	2003	2004	2005	2006	2007	2008	2009
GDP growth (%)	8.2	5.4	4.6	3.5	9.6	6.6	5.8	5.3	5.7	6.0	6.7
Oil sector growth(%)	6.6	11.1	5.2	−5.2	23.9	3.3	−1.7	−3.7	−5.9	−6.2	−1.3
Non-oil sector growth (%)	8.6	4.4	2.9	4.5	5.2	7.8	8.4	9.5	9.2	9.0	8.3
External reserves (% of GDP)	NA	NA	NA	NA	7.7	11.4	24.4	36.5	42.6	52.99	62.48
External debt/GDP	106.5	64.9	57.3	72.1	61.1	84.5	69.2	7.4	4.0	17.5	9.28
Domestic debt/GDP	31.6	32.2	36.6	26.1	28.6	25.3	20.8	18.6	19.2	15.23	12.85
Overall BOP/GDP	−2.1	6.9	0.5	−10.3	-2.3	5.2	10.5	12.7	1.4	8.02	9.12
Inflation rate (%)	7.5	6.9	18.9	12.9	22.7	15	17.9	8.2	5.9	11.6	11.5
Average official exchange rate (Naira/US$)	7.9	101.7	111.9	121	127.8	132.8	132.9	128.5	127.4	139.27	142.89

Adopted from MDGS report, (2010).

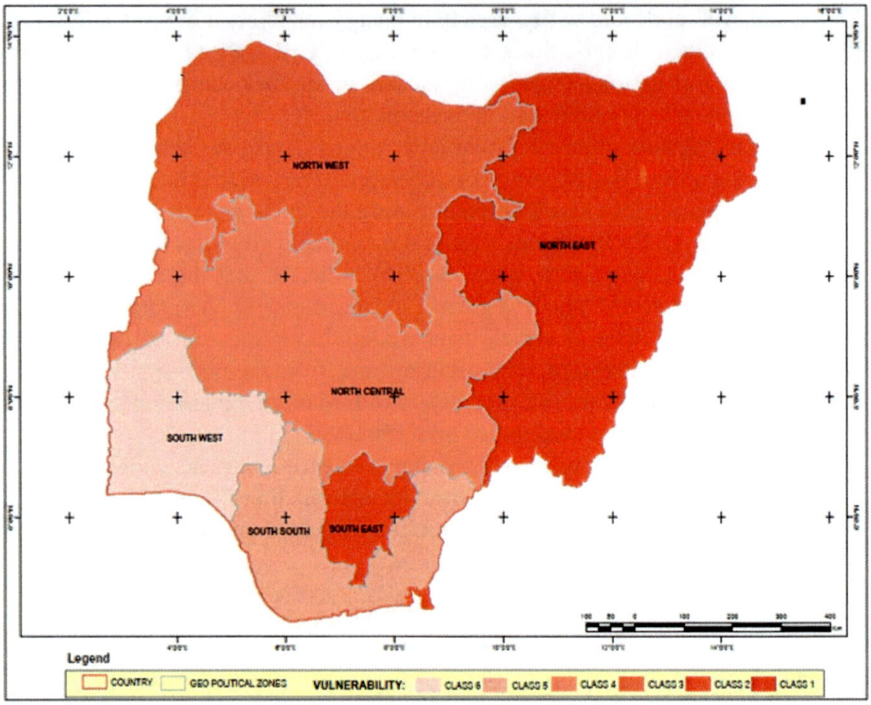

Map 3.1 Spatial variation in relative climate change vulnerability. *Source* Second National Communication, (2014)

Large-scale population displacement; drought, desertification, flood, food insecurity; low yields in agricultural and fishing industries.[19] Worst still, climate change will seriously affect agricultural activities of the people given that only 5% of the population have access to improve and climate resilient seeds. Still, the incidence of floods which is projected to rise will become a perennial occurrence surpassing the 2012 floods.[20]

Floods in Nigeria are recurrent annual climate events. There is, in fact, a long history of floods.[21] Various incidences of floods have occurred in the country's geological history. Devastating floods have hit different cities and regions recorded in previous studies. These floods are known to have paralyzed economic activities in some parts rendering many people homeless, destroying roads and bridges ultimately affecting agriculture and human development.[22] The 2012 flood seriously affected Benue State and Kogi State since the people are largely farmers. Perhaps more important, River Benue and River Niger provides the means for agricultural and economic activities of the people who use the different streams and rivers for their farming activities. The rivers have significant effect on agriculture, land use encouraging irrigation and cultivation along the river Benue Valley.

The 2012 floods, however, changed this notion of the river as providing the means for existence to a source of woes. Evidently, the flood is the most devastating ever experienced in the country since 1956. The regrettable fact is that adaptation strategies are not region specific nor do they take into consideration the cultural, historic and bio-geophysical specificities and attributes of each region.[23] To resolve this will require new strategies geared towards region specific challenges that are gender sensitive and participatory if the people must cope with the mutually reinforcing challenges of extreme hunger and climate change. Climate change will, undoubtedly, impede progress towards the Sustainable Development Goals (SDGs) with the objective of ending global poverty by 2030. It will also have severe ramification for the Vision 20:2020 and the ERGP long-term development agenda of government. More importantly, it will affect the outcome of decarbonisation making it difficult to build a low carbon and climate resilient economy.[24]

Nevertheless, economic transformation requires burning dead or living carbon. Burning dead carbon such as petroleum fuelled extensive discoveries engendering remarkable improvements in living conditions globally. For this reason, carbon emission is attractive to most developing economies because energy is an integral part of economic development. In a developing economy like Nigeria, burning dead carbon provides households and industries with much needed energy. Energy generation has a key role in advancing government's transformational programmes crucial for national wealth creation, employment and improved standard of living. Like other developing economies, Nigeria keyed into this free market assumption of carbon-led growth being necessary precondition for the transformation to a middle-income nation. Along this reasoning, the Vision 20:2020 blueprint is target at transforming the economy using abundant fossil fuel energy. The simple argument is to maintain the pace of economic growth at the rate of 6–9% GDP annually; it will require increases in the extraction of fossil fuels to pave way for economic development.

Since the nation is lock into a fossil-dependent economy, they recorded impressive rise in GDP (before the economic recession of 2020). It is trite to state here that remarkable transformation recorded in the last decade is a product of GHG emission. That is the truth! In addition to biomass extraction, GHG emissions from gas flaring created adverse pollution of the ecosphere.[25] According to the International Energy Agency (IEA), in order to have a 80% chance of maintaining the warming limit through to 2050 the world needs an additional $36 trillion in clean energy, or nearly 15 times Africa's nominal GDP.[26] This will place severe burden on Nigeria's Vision 20:2020 since it will require additional financial commitment. It will further mean leaving dead carbon on the ground if the country is to meet its share of maintaining the carbon uptake at 450 ppm considered safe to avert dangerous climate change.

Furthermore, the pollution-abatement strategy crafted at UNFCCC insists different nations embark on a less carbon intensive pathway thus requiring a further squeeze on the fossil fuels exploitation. Placing a squeeze on fossil production is, to put it mildly, a big task. The general sentiment within Nigeria has been one of feeding citizens which is considered a bigger priority for government than focusing on the shift to clean energy, even if the benefits and consequences of not doing so are apparent. This is a reflection of Nigeria's sentiment encapsulated in the National

Report to Rio +20 in 2012.[27] The former President of the African Development Bank (AfDB), Donald Kaberuka captures Nigeria's sentiment when he bemoaned the imposition of irrational emission targets on Africa. Kaberuka took this stance when the AfDB's decision to continue financing power plants that use dirtier coal was questioned.

According to Kaberuka, the hypocrisy of western governments who funded their industrialisation using fossil fuels made them now to ask African countries not to develop dams and cannot develop coal but is meant to rely on very expensive renewables. Nigeria like most other African countries took this line of reasoning and did not listen to the world body because in every single African country the biggest impediment to economic growth is energy poverty. Today, most countries are travelling on the renewable energy road hopeful it will rein in perennial energy poverty crisis in addition to improving their economic circumstances. This is a tall ask but eventually a realisable aspiration.

The Third Industrial Revolution is now a global reality but it is not void of problem for Nigeria's economic wellbeing. Not only will it change oil sector's role as the signpost of GDP growth, could reinforce assertion of oil being the poster child of all that is wrong with the economy. Arguably, steps of improving stabilization of the biosphere carries with it additional burden of running fossil fuel off the energy ladder. Government faces the economically difficult task of acquiescing to the new energy regime or maintains the ecologically negative role of a country with outmoded reliance on petroleum. The plug-in car revolution otherwise known as the third industrial revolution is fast catching on with major fossil economies laying out their FES. This FES represents ambitious action plans to radically reorder production and consumption of fossil fuels with the goal of sustainably using resources.

Given its immense co-benefits, investments in electric mobility have increased in the process has changed the values in customer expectation. Moreover, this cultural shift hinges on plug-in cars' role considered a part of modern urban lifestyle, it is equally based on the need to live sustainably in an eco-efficient manner. Generally, the EVs revolution is part of another landmark event targeted at tackling anthropogenic GHG emissions. The growing urgency attached to the decarbonization process in the form of the proliferation of EVs technology is the fallout of another equally outstanding event: The COP21. This much we have stated severally. COP21 opened a new vista in climate governance thereby ensuring transition from political indecisiveness to laying out a concrete climate plan of action. A remarkable outcome of the conference is of some countries laying out their FES. These countries have agreed to embark on a low carbon trajectory using the transportation and energy generation infrastructure as its spring board for reducing the high uptake of carbon dioxide.

So, one can say that fossil fuel consumption is gaining less footing but it has not been totally eradicated due to its strong influence. Note, however, it is highly unlikely fossil fuels will run out in the near future- especially if future technological change makes major new sources like oil shale and methane hydrates commercially viable. Some have argued that alternative sources of clean energy like solar and wind power used for generating electricity and fueling electric vehicles have seen substantial progress in their cost reduction[28] (at least in the short and middle-term) they are unlikely to play a major role in base-load electrical capacity or in replacing petroleum-fueled internal combustion engines. Moreover, carbon consumption and production are necessary components of the modern age even though it is now man-made climate change major causation.[29]

As noted elsewhere, part of the solution proffered for mitigating climate risk at international level insists on keeping fossil fuels on the ground[30] as a consequence would in the long run reduce the carbon dioxide (CO_2) level in the atmosphere to below two degrees Celsius (2 °C) threshold considered safe for human existence by Copenhagen Accord and later the Paris Agreement. With the Paris agreement becoming the yardstick to measure commitment on reducing carbon upload, the stage has been set for major fossil-based economies to eliminate fossil fuel from their energy mix. Today, the exponential explosion of EVs production together with other renewable energy technologies are considered the emerging face of energy efficiency because it strives to correct harm posed the global economy from the Second Industrial Revolution. Critical to this revolution is the inevitable decrease in production of Petro-powered cars noted to have added adverse carbon load which is now exacerbating the build-up of atmospheric GHG. The global decarbonization is, indeed, underway. Current strategy to reduce GHG emission represents progressive, ambitious action plan capable of not only radically reordering fossil fuels production and consumption, it will foster a sustainable energy future through the development of other viable climate-proof alternatives.

This low carbon technological shift may be detrimental to oil producing nations' economic growth. For this reason, some countries such as Saudi Arabia have developed their roadmap to a post-fossil fuel future. Till date Nigeria has not fashioned any concrete plan for the post-fossil-fuel society to take root. The nation is still highly dependent on production and exportation of crude oil for domestic use and export.[31] In the country, the oil and gas sector provide over 90% of the country's foreign exchange earnings and about 80% of all budgetary revenue.[32] Without any significant plan for the future, the Third Industrial Revolution could totally wipe out gains from fossil-led growth, however.

Pressures from the declining cost of solar and wind power could also shift development of renewables from being policy-driven to market driven. Cost competitiveness of these technologies could trigger a radical shift similar to the GSM revolution. A similar process maybe expected when used (second-hand) electric vehicles are readily available in the country. Prospect for this is still far from now. Furthermore, development in lithium-ion storage is improving globally with the number of hours of power storage; pace of uptake of the technology and market changes are high. Given this, different societies are gradually awakening to the renewable revolution while others within Africa like Rwanda, Ethiopia and Kenya are toeing the low carbon trajectory for the obvious reason.

With various societies now embracing the renewable energy option, economic decision making will be put under severe stress. The ability of governments to tackle the ongoing cultural shift will hinge on how these policies support efficient markets and deployment of new technologies, if we are to avoid having stranded assets. Energy security in the near future is a function of government's ability to avoid locking onto stranded technological choices in opulent pursuit of stable and efficient power from renewables. For now, there is no such luck in the horizon rather the country is amassing a plethora of stranded assets with older fossil electricity generation capacity more visible in the national supply energy mix. Deployment of alternative energy solutions together with the decarbonization agenda will render most gas and petroleum infrastructures redundant necessitating the formulation of policy solutions on economic diversification in order to incentivize supply shift towards cleaner energy infrastructure.

Nigeria's Carbon Catch-22

Nigeria like most oil producing states is confronted with the task of either embarking on a low carbon pathway or continues with the business-as-usual fossil-led growth path hopeful her ecologically harmful emissions will not undermine development. In this, the country is with the hard choice of either embracing a clean energy path by rejecting 'dirty' or 'brown' resource intensive path or proceed with the BAU 'brown' economy. From the low carbon plan of action in the National Renewable Energy Action Plan (NREAP), there is high hope for permanently reducing our adverse carbon upload. Despite this, the low carbon journey has series of endogenous and exogenous related crises with both economic and political underpinnings bedevilling it. The option available to government is either toeing a low carbon emission route or finds a way of overcoming the economic stress spurred by the 2012 floods and 2015 and 2020 oil shock which threw the country into doldrums of economic uncertainties.

Unarguably, any country at the threshold of coming out of economic depression must take urgent and hard options of either penalizing fossil fuel or bearing the geophysical consequences of their inaction. Though avenues for achieving the goals of decarbonization are many but the real choice is limit thus making the plan to reach the top of carbon emission reduction's mountain a herculean issue of monstrous proportion. In the light of this, there is a sense of *déjà vu* in addressing the dilemma. Nigeria's carbon impasse can also be labelled a carbon catch- 22 situation. What is Nigeria's carbon catch-22 and why are we using the term to describe the country carbon complexities? One may ask.

A carbon catch-22 is generally a paradoxical situation where a country in need of growing her economy uses their abundant resources in a manner such that goods and services required for human development are provided. Still, this same country is expected to reduce the carbon intensity of the economy using these same resources in a manner that it does not add to their huge carbon debt. So, the elephant in Nigeria's carbon budgeting room is how to overcome dilemma of upholding environmental protection commitment while pursuing a steady economic development path. Effectively, reprioritization of development plan to ensure the Paris commitments are realised while also fleshing out an equitable growth path is Nigeria's albatross. Rebalancing consumption and production of carbon does not totally mean the pace of economic development gaining traction would totally wear away, however.

Far from it! Rather, it amplifies need for charting a middle course between these competing issues.

Again, a trade–off between environmental pressure and social services provisioning of government is required. Accordingly, the main talking point here is whether the vast population of depraved Nigerians (Nigeria overtook India as the poverty capital of the world in 2018) will consent to imbibing a pro-nature stance which leads to the reduction of their immediate social needs. Resolving competing interests between people-centric and eco-centric development is now the ecological *cue de sac* bedevilling Nigeria's effort to forge a carbon neutral path.

Diverse threats to the corporate wellbeing plague Nigeria but three major interrelated constraints have far reaching consequences for the ship of the state. These three closely connected issues are: COVID-19/dwindling economic fortunes, energy poverty and inability to address the upsurge of climate-related risks. These cross-cutting issues have one thing in common: they are principally the offshoot of natural factors as well as have human induced causation. As already noted elsewhere, the first impediment is the outcome of Coronavirus of Wuhan while the second is outcrop of several interconnected issues paramount amongst which is the proliferation of EVs and oil glut. The third is a function of economic and environmental limitations also linked to the carbon intensity of the global economy. Collectively, these issues do not only undermine development but affect the political and economic wellbeing of the state. To address these problems, require filling gaps created by these closely-knit issues. Putting Nigeria's carbon-based economy back on track of symbiotic relationship with nature is one of the solutions advanced at the international level.

Changing Nigeria's carbon footprint also calls for the elimination of extraction and consumption of fossil fuels. Amongst the two options, the latter seems more easily accessible with the co-benefit of placing the economy on a broad-based path to sustained growth in a manner that it does not exert serious pressures on a single resource. Still, this is the most complex path equally linked to another endogenous variable- the Dutch Disease. Moreover, the petroleum sector holds great importance to the economic health noted to have provided the people with immense economic benefits while other sectors are largely underutilized. Despite promises of oil elevating the standard of living of the people, it is not totally without consequences. The consequences of reliance on a single commodity will come when the petroleum bubble bust. The day of reckoning is at hand since oil mortality has come to stay. Indeed, the day is now. Due to lack of saving, there is now no plan for the proverbial rainy day. Since there is no saving, without a doubt, there is no plan for a future without oil. To put the matter more forcefully: Despite all the promises of a prosperous future based on exploitation of crude oil, it has not met Nigerian expectations. A major impediment is the want of concrete plan of action on economic diversification.

What has emerged is that Nigeria is not rooted in long-term disciplined plan for disengaging from fossil fuel-led growth. In the real sense, one may draw conclusion from the proliferation of fossil-fuel power generation that the country seems more entangled with fossil fuel-led growth with understandably insignificant presence in other alternatives. The result is food and energy poverty exists side by side with stupendous wealth but they are as co-equal forces. Today, the massive oil wealth generated has not translated to economic wellbeing nor has it changed the standard of living of the people as it ought to. Though some remarkable changes took place before the 2014 national account statistics, it was not with the much-desired improvement in living conditions of the poor. Instead of improving the standard of living it is rather plummeting into the deepest ravine imaginable. The hard-won improvement in GDP took further bashing from blows unleashed by the 2015 oil depression which plunged the poor into further misery. Because of this, the incidence of hunger which hitherto had decreased has now become a major killer (literally and figuratively) making the poor to eke a living from hiring out their labour to the super rich.

Even the superrich will not be spared from the onslaught of this economic downturn. Prior to the depression many rich Nigerians had fleet of cars. But with the oil stress of 2016 some had recourse using only one utility car instead of their Petro guzzling posh SUVs. This is a positive development in a polity not accustomed cutting corners or placing barriers on squandering their ill-gotten wealth. Another positive development isa large population who consume locally produced rice exist since imported rice fell out of the plate of the middle-income earners. The positive effect is most Nigerians now patronize the local rice variety while development planners now look inward for local remedies for imported brands to reduce the large amount of money used to import food.

Across the globe, the EV revolution is catching on with its tentacles dangling like a Sword of Damocles over the oil sector's head. A discernible future is gradually emerging from the haze of poisonous fumes of fossil fuels because pollution-abatement solutions have taken their toll on the oil sector. Consequently, it has moved consumption and production away from the established contour to an uncertain economic terrain. As most societies continue to grapple with these changes, the ship of the renewable energy sector is set to ratchet up gains amassed during the ill-wind of the economic shock is about to sail. This ship to the post-fossil fuel future is ready to sail with or without oil producing nations.

The end results of this is that the large cache of arms gathered by oil kingpins to stifle the renewable energy transition will not suffice instead will like the Titanic cause it to capsize. Fossil fuel capitalists' have in their haste initiated some green measures to rescue the sinking ship of fossil fuels but this is nothing more than 'bailing out a few buckets of water as the Titanic sinks'. To them, decoupling nature's exhaustion from economic growth is not possible since capital sees ecological and economic sustainability to be mutually contradictory. Despite this defective ecological reasoning, clean energy will still gain more traction globally. In Nigeria and other oil producing nations, it will cause their fossil fuel infrastructures to become abandoned and stranded assets when these chronologically non-functional infrastructures have become moribund.

Impact of Oil on Nigerian Economy

The oil industry due to its control of the source of national income has an overbearing influence over conflict resolution mechanisms. Who control of instruments of coercion is material to laws governing oil operations. Unarguably, instruments of coercion has been used to private end hence is causing distributional crisis. However, distributional, allocative and externalities crises resulting from policy deficit are recipes for armed conflicts with instruments of governance used at behest of those who dominate the oil sector and not by the host communities. Overcoming this crisis demands fundamental reorganization of the way of doing business.

Ever since the discovery of petroleum in 1956 Nigeria recorded mixed blessing in the socio-economic and political spheres. Nigeria's economic fortune is highly dependent on the rhythmic movement of oil. This is because oil considered the heartbeat of the nation could be also Nigeria's Achilles heel. The large deposits of fossil fuel which created modern wealth also caused inequality in some societies (most especially in the south-south geopolitical zone). Though large deposits of crude oil engendered economic development, it also created an army of rent seekers and a mass of underprivileged class at the other end of the economic spectrum. With crude oil-led growth there is the co-habiting extreme of both abject poverty and stupendous wealth in the south-south region. Inequality in Nigerian society is an eloquent testimony of allocative crisis equally linked to distributive crisis wherein the distribution of oil wealth is skew towards one class. This explains why less than one percent of the population now control over 70% of the wealth with the bottom 25% living below the poverty line.

Fossil fuel provides immense wealth but also exerts serious environmental pressures on the fragile ecosystem. No doubt, the price tag of Nigeria's stupendous wealth from oil is the ill-health of the environment and the bifurcation of different societies. Nigeria's economic development which resulted in a stable rate of 7% annual growth rate brought with it untold hardship for a clear majority of the people. So, a huge carbon debt is owed future generation for manifold fossil-led development we have recorded. This carbon debt is outcome of the negative footprint of crude oil extraction processes, transportation and consumption because all these activities come with a huge cost. The environmental cost of our improvement in living conditions is the massive spills from oil extraction and transportation which is changing not only the shorelines but also the marine ecosystems.

This source of wealth facilitated growth of other sectors, however. The boom cycle created further wealth that should have been ploughed back into the economy but only succeeded in reversing the gains made in GDP growth. In commenting on the development shortfall resulting from the oil curse Paul Collier decried Nigeria's attitude of not planning for the future thus: 'At times of high oil prices the revenue has enabled both ordinary citizens and the state to achieve higher levels of consumption, these levels have not proved sustainable: windfalls have not been converted into permanently higher levels of income. The disastrous failure of the previous oil boom of 1974–86 is often misunderstood in Nigeria. During the boom consumption, both public and private, was high, and so living standards were temporarily high. The catastrophic failure of public policy was that the windfall has not transformed into a higher level of sustainable consumption. On the contrary, policy during Nigeria's oil windfall ensured future consumption would be higher.[33] This, according to Paul Collier (2008), is because during boom periods the country consumed all the revenue instead of investing in productive assets and when the petroleum 'bust' occurred government borrowed very heavily to finance consumption.[34]

In the following sections we shall explore how the country got to the current situation by examining changes in the economy, agriculture and policies of government over the years. This exposition will enable an appreciation of the gaps in policy formulation and implementation thus arming us with the information to proffer solutions for an equitable transition to the post-carbon future. To appreciate where we are as a nation, we must understand how we got to where we are today. Carrying out such analysis is a delicate matter prone to various economic and political interpretations but there is no 'royal road' to Nigeria's economic emancipation neither are there quick fixes. We shall presently examine some of the issues with overwhelming influence on the country:

I. Policy Interventions—Like most developing nations Nigeria never had shortage of well thought out policies on improving the living standard of the people. Before independence, agricultural sector was the mainstay of the economy. The country was a major agricultural exporter with groundnut, cocoa, cotton, rubber, palm oil and hides and skin contributing over 60% of the GDP. Different regions specialized in crops they had comparative advantage. There were diversification based on comparative advantage such that groundnuts where exported from the Northern parts of the country, cocoa and rubber from the Western region and palm oil from the Eastern region. All these created a robust export oriented economy attuned to the agricultural activities of the people. Little attention was paid a single commodity at the time with all regions in a healthy competition. It was therefore not surprising that the colonial policy aimed at expropriation of surplus commodity based on the agricultural policy of 1946 was adopted by the regions at independence. Likewise, Nigeria Cooperative Ordinance Act of 1935 which led to the establishment of commodity marketing boards in 1947 further created vital link between farmers and the markets.

Nigeria was already oil exporting nation exporting at independence. About 415,000 barrels per day of crude oil was exported by 1966. With revenue accruing from oil, the planners of the First National Development plan, 1962–1968 did not make serious attempt to develop the agriculture sector. No conscious effort was made to encourage private investment in agriculture. At that time public investments were skewed towards elite preference of large scale agriculture prone to obnoxious values. As if this harm done the economy was inconsequential, the Second National Development plan, 1970–1974 replicated errors of the First National Development Plan. Burgeoned by petrodollar resulting from the oil boom agriculture lost its role as the major source of revenue which was 75.9% in 1965 but contributed only 2.4% in 1980 while petroleum appreciated from 2.7% in 1960, to 73.7% in 1971 and 96.1% in 1980.[35]

Thereafter, the political space has become a graveyard of policies. These policies includes: The National Development Plan (First—Fourth) NEEDS, Seven-point Agenda, Agriculture Transformation Agenda, vision 20:2020, MDGs, ERGP and a host of other medium-term instruments. With dissimilar policies competing for attention there was lack of synergy between different policies and actions of government. The amount of inconsistency resulting from the want of political continuity remains a major drawback. Not that there are no policies formulated for the good of the environment rather the inconsistency in the policies created apathy within the citizens and the organized private sector towards environmental protection. In 2004, the NEEDs blueprint attempted to correct some of the inadequacies observed in the previous policy instruments. Despite this was entangled in the political horse trade between vested interests.

On assumption of office the Yar' Adua Administration launched a seven–point Agenda for substantial transformation linked with macroeconomic changes. The former President Goodluck Ebele Jonathan's Administration on assuming office took up the mantle from where his predecessor stopped. Whereas President Muhammad Buhari's Administration is more inclined towards welfare centred principles interlaced with market economy tenets. His administration accurately assumed that microeconomic transformations will usher in a dawn of industrial economy. The main instrument to drive this transformation is the ERGP. At its basic level, the ERGP is a medium term policy instrument targeted at both macroeconomic and microeconomic change interspersed with strong agricultural development backbone.

II. The Economy—Nigeria sacrificed her environmental quality to expand the manufacturing, financial service, communication, technology and agricultural sectors. The country is an emerging economy, the 23rd largest economy in the world in terms of Purchasing Power Parity. The nation had an annual growth rate of 6–7% before the 2015 recession, but since then is with a GDP growth of −2.1% while inflation stood at 9% with a population of about 200 million rising at an average rate of 2.5%. This means the country will exponentially grow to becoming the 4th most populated country in the world by 2050. With an economy heavily reliant on petroleum where petroleum products account for $93.01 billion while imports stood at $52.79 billion, the economy can be said to be growing with the GDP PPP tripling from $170 billion in 2000 to $451 billion in 2012, whilst the GDP per capita doubled from $1400 per person in 2000 to an estimated $2,800 per person in 2012. (See: Fig. 3.1).

Going by this, the country is a middle-income economy but is not totally free from the dilemma of overcoming economic challenges peculiar to low income countries. Bill Gates explicitly points this out to the Federal Government when he was invited to the National Economic Council held at the Presidential Villa, Abuja in March 22, 2018. According to the multi-billionaire, the trajectory of Nigeria's GDP growth should go straight up (Fig. 3.1) except for the recent recession. Regrettably, the nation is still trapped by the low middle-income nature such that there is no discernible difference with poor countries. As he puts it: 'From the point of view of the quality of life, much of Nigeria still looks like a low-income country. Let me give a few examples. In upper middle-income countries, the average life expectancy is 75 years.

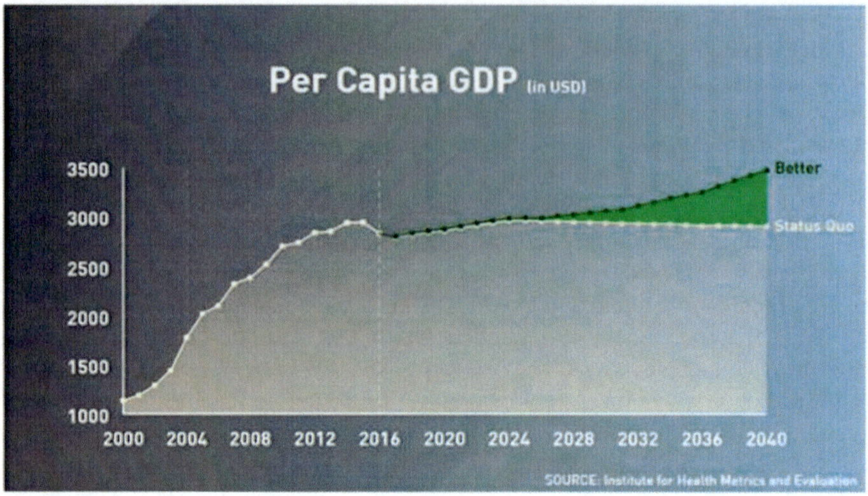

Fig. 3.1 Projected per capita GDP USD. *Source* Bill Gates (2018)[35]

In lower middle-income countries, it's 68. In low-income countries, it's 62. In Nigeria, it is lower still: just 53 years. Nigeria is one of the most dangerous places in the world to give birth, with the fourth worst maternal mortality rate in the world, ahead of only Sierra Leone, Central African Republic, and Chad.[36]

On the basis of this, Nigeria can be classed a low-income country in the league of countries like Chad, Niger and Sierra Leone. To the government and people of Nigeria this analysis is a painful truth. This is very hard to swallow most especially when it comes from one who is the highest individual investor in Nigerian health sector. The Vice-President's hasty rejoinder attempted to correct the impression created. On the contrary, his rebuff created new problems rather than resolved Bill Gates' assertion. Despite this effort much harm had already been done. The unfathomable fact still remains that the country despite all positive data is at the nadir of low-income countries in terms of quality of life. Though oil revenues contribute 2/3 of the country's revenues, it only contributes about 9% to the GDP. Nigeria according to the World Bank had the highest growth rate in the world by 2014 at annual growth rate of 7.4%. The economy, however, took a sudden downturn with the 2015–2016 oil shocks.

Though Nigeria came out of the recession after two tortuous years of economic hardship but coming out of the recession was outcrop of oil price appreciation backed by solid performance from the non-oil sector. It is well known fact that agricultural sector is now the rising star of the Nigeria economy and this rise is an outcrop of sound agricultural policy. Agricultural policy of the administration placed serious emphasis on local production of food crops such as rice and advanced loans to the farmers to record the impressive crop yields we witnessed. The Anchor Borrower Programme which is flagship project of the administration gave smallholder farmers access to loans and inputs, for example. With the recovery from recession government invested heavily in infrastructural development hopeful it would engender endogenous growth. Nigeria's solid performance is, however, heavily dependent on hydrocarbon production.

The 2020 COVID-19 crisis changed the positive growth rate recorded in the country. Given that the country did not plan for oil price slump, it felt the full weight of the economic and health challenges. Decoupling fossil fuels from economic development is supposed to be the outlook of government in the pre-COVID-19 era. Characteristic of previous administrations making provision for economic meltdown by way of economic diversification was not on the front burner of policy. To a large extent, it was relegated to the background since such an action will reverse gains in economic progress recorded after the 2015 recession.

With the looming danger of eliminating fossil fuels from the global energy mix, long-term changes in the economic circumstance of the nation are inevitable. It may seem oil's potential of raising Nigeria from intermittent economic shock is a direct threat to solution for ending ecological pressures. Even though Nigeria is under severe pressure to fine a viable solution to generate revenue while exiting 'dirty' or 'brown' economy such way out is construed to undermining oil crucial role. Little surprising then, that glowing tribute was paid economic diversification with insignificant actions commensurate to the utterance to actualise this goal.

It is not all sorry sight where oil economy is concerned. The economy has had a spectacular history with diverse sectors holding their sway in the past. At the onset, the economy was heavily dependent on agriculture. In the colonial era, agriculture was the dominant source of income contributing more than 60% of the GDP. The discovery of oil gradually changed the economic fortunes of the country. The oil windfall instead transforming the agricultural sector wound down the agricultural base in the process turning the country into a monoculture economy. The weakened agriculture sector took further bashing in 1990s with the change in economic structure from dependence on agriculture to a service-based economy.

Dr. Yemi kale, the Director General of Nigeria Bureau of Statistics (NBS) in a paper titled 'Pushing Nigeria's Economic Diversification Forward: Issues & Options delivered to the 7th Stanbic IBTC Bank Investors Conference noted how the price level of oil itself is not a problem, and that a bigger challenge lies with oil price volatility.[37] As a result of uncertainties of the oil market agriculture which accounted for 35% GDP slumped to 22%. Similarly, industry including crude oil and gas production which used to be 36% of GDP fell to 25% in 2016. Whilst services that used to be 39% of GDP was over 50%.

Nigeria's inability to translate development plan into physical change is to blame. As a nation reliant on oil for revenue, implicit multiplier effects for the entire economy have been most staggering and pervasive, so said Dr Kale. Although many would have noticed imbalances and strains that were beginning to appear in the economy stemming from gradual decline in price of the country's main foreign exchange earner, very few could have anticipated the extent and severity of the economic and financial storm that has subsequently enveloped markets. Despite the unprecedented number and scale of the policy intervention measures introduced, systemic pressures have not fully abated.

As of late, COVID-19 crisis has decelerated gains of 2019 seeing that the country recorded negative growth due to oil prices reaching an all-time low only recorded in the last 30 years. Couple with a high debt portfolio, the country was headed for further economic downturn. In the mid 2020, confidence in the economy and the financial markets gradually developed but the economy is thus far not reverted to 2019 growth rate. Just as Dr Kale reported in 2016, the economic growth rates have slowed from a position where Nigeria was one of the fastest growing economies in the world to one of the slowest in April 2020. History is repeating itself in 2020.

Though the country made significant progress between 2010 and 2014, the economy took a nose dive in 2020 due to policy shortfall of not planning for oil shocks. Consequently, the oil shock resulted in the slump in prices from 66% which was about $114/barrel recorded in June 2014, to as low as $31.4 as at 2016; which sold $15 in April, 2020 with the market picking up to $43 in July, 2020. Nigeria's vulnerability to external shocks together with uncertainties of the parallel market made development planning a complex issue. Economic lockdown of the global economy resulting from COVID-19 hazards in addition to the Russian and Saudi Arabian oil price wars of 2020 boomeranged on the global energy market snowballing into a mega crisis.

The reliance on oil had a multiplier effect of plunging the nation into one of the worst recessions in the annals of the country's economic history. Accordingly, confidence in the financial market also drastically declined. The end result is rapid capital flight and consequential reduction in FDI. The economic growth rate projected to stabilize around 7% was the first major victim of the recession. The growth rate which slowed down to about 3% in 2015 while NSE AI of 2016 was at 23,171.07 points recorded in 2010 which is about half what it was at the peak in 2014 but in 2020 was in a negative territory. With the decline in revenue generated to meet developmental goals, government's 2020 budget targeted at infrastructural development became an extremely difficult task. With this deterioration, the imperative for a blueprint with economic recovery at heart was high. To stem further weakening of the economy a committee for a post-ERGP blueprinted was constituted by the President Muhammadu Buhari targeted at creating jobs and injecting capital to stem the downward spiral.

III. Agricultural Sector—Over the years, agriculture's role as tool for national economic development has waned. In its place public expenditure was elevated to the highest summit of the policy pyramid while the real groundnut pyramids of Kano disappeared never again to appear on Nigerian agricultural horizon. In place of investing in people through agriculture, the planners of the First, Second and Third National Development Plans made insignificant overtures to agriculture. Though efforts were made to promote the productive sector, it was bias towards capital intensive large-scale agricultural projects. It is therefore not surprising that a decline in export of commodities was recorded.

In aftermath of the World Bank imposed Structural Adjustment Programme (SAP), agricultural policy toed the line of deregulation. Commodity Boards- the hydra headed monsters were hurriedly dismantled paving way for farmers to have direct access to markets. However, deregulation brought with it additional burden of paying large food import bills. New polices targeted at improving the wellbeing of smallholder farmers saw government instituting the National Agricultural Land Development Authority (1991–1999), the Presidential Initiative on Cocoa, Cassava, Rice, Livestock, Fisheries and Vegetables (1999–2007). Though these initiatives aim to increase agricultural yields through supply of agricultural inputs and mechanization but disconnect exist between these government roadmaps and policy implementation.

It would not be unreasonable to state here that agricultural progress sought by various blueprints to resolve dwindling yields in the agricultural sector produced immense wealth for a select few. An efficient system of improving agricultural yields seems to have eluded the country which is linked to the 'resource curse' and 'Dutch disease'. The 'resource curse', without a doubt, is affecting the 'black gold' but it has also boomeranged into the agricultural sector decline. This explains why a country with immense agricultural potentials at independence is now at the threshold of agricultural doom.

It is pertinent to underscore that the degeneration within agricultural sector is inversely relational with the 'boom' and 'bust' cycles of the oil industry. Evidences from 1976, 1982, 2007/2008, 2015/2016 and 2020 show that the energy crises are closely related with periods of food crises in Nigeria. One would expect after many years of series of oil related crises, the economic planners of the country should have gotten their acts right to plan for these cyclical events. In its place the boom period created new incentives for accumulation of surplus for reinvestment. The boom gave rise to renewed impetus for profligacy in investing in infrastructural development. Though some administrations may have good intentions but good intension does not engender development, neither does it foist climate solvency. For climate solvency development policy must be driven by well-equipped bureaucrats who then can propel economy towards new heights. For now, there is no such luck in the horizon.

The policy thrust of this administration changed towards higher policy implementation and service provision. Minimal successes are recorded in some sectors to tally with the goal driven agenda of government with agriculture functioning as the engine of growth. So, the agricultural sector is undergoing reforms through value chain transformational policies. A key factor driving this change is the Agricultural Promotion Policy (APP). APP strategy aims at improving crops yield with additional benefit of boosting resilience through preparing communities to deal with resource scarcity and extreme event. Likewise, the Nigerian Agricultural Resilience Framework (NARF) sets a plan of action for climate change risks adaptation. To improve food yields the APP identified Climate-Smart-Agriculture (CSA) to be the solution to agricultural-related deficiencies. The CSA specifically seeks to address climate change while fostering food security based on efficiency, productivity motivational adaptation. Primarily, it is meant to act as the building block for transforming the agricultural sector. The major objective is to achieve green growth using ecologically sound principles to ensure sustainable food production.

To achieve sustainable structural transformation of the agriculture value chains, major stakeholders must be carried along. Agriculture is an employer of 80% of the

Nigerian working population. It requires additional policy for the informal sector, technology incentives and capacity building to absorb teeming population of unemployed Nigerians (33.6% of the population equivalent of 39.4 m in 2020) into gainful employment capable of raising the revenue base through instrumentality of taxes. If the informal sector is mobilized together with the unemployed, they will contribute to national development via taxes in so doing bring about an inclusive growth. It is quite unfortunate that agriculture only contributes 0.11% of total VAT, thus far 18.45% to GDP.

A more robust farming approach should redirect productive energies towards more radical agricultural systems. This entails using technology to transform from subsistent farming to industrial agriculture, in so doing overcome uncertainties associated with oil price volatility. There are new structures with the overarching goal of revamping the farming systems, hence the significant progress made in reducing rice imports. Still, there is room for improvement. It may seem that transformation of the rice producing sector is one incidence in isolated instance without additional benefit of a trickle down to other facets of the agricultural value chain.

The budgetary allocation to FMARD has been fluctuating over the past years with 2015 having a lean allocation while 2013 had the largest allocation. Nevertheless, there is a trend towards agriculture in recent budgets even though huge gaps exist with allocation to recurrent vote and capital vote (Table 3.2). From 2014, government showed commitment towards agriculture evidenced in the increased budgetary allocation.

The Anchor Borrower's Programme raised the agricultural yields even though the percentage of fund allocated to this (Table 3.1) shows commitment to infrastructural development with fossil fuel power generation the most favoured. This is quite understandable. On balance, power dynamics at play does not really favour agriculture. Government's total control of agriculture and involvement is necessary for ensuring people have adequate food. The present attempt to improve agricultural yields is commendable for ending near-term hunger but it will not meet SDGS #1 of ending global hunger by 2030 neither did it stem COVID-19 imposed food insecurity in 2020. SDG#1 is only attainable when funds diverted to private accounts of politicians are judiciously invested in agricultural projects.

Table 3.2 Federal allocations to agriculture 2013–2016

Year	Overall budget	Allocation to agriculture	Recurrent vote	% of Recurrent vote	Capital vote	% of capital vote
2016	6,060,677,358,227	29,632,584,416	632,584,416	39.53	46,173,963,859	60.47
2015	4,493,363,957,158	40,659,020,717	869,020,717	78.38	8,790,000,000	21.62
2014	4,075,190,000,000	66,644,675,939	493,503,356	47.25	35,551,172,583	52.75
2013	4,987,220,425,601	83,762,937,710	31,115,066,282	39.53	50,647,871,428	60.47

Source Budget Office of the Federation (2013–2016) cited in CSJ, 2016

Ending hunger by 2030 seems farfetched but is achievable through reallocation of over 80% of fund committed in 2016–2019 budgets to capital vote to agriculture. When such a remarkable step is taken, the journey towards ending hunger is half won. The other half hinges on the 'Nigerian factor' of derailing any threat to shareholders returns. With food (rice and tomatoes) importers' political visibility, the promise of ending hunger is nothing more than an incremental tweak at the problem, while the real issue remains unchanged. Because, they are incremental tweaks in isolated circumstances when one target is realised in one window, in the next circle a more devastating crack is reinvented inevitably negating earlier gains made in the last cycle. Nigeria must take advantage of the current COVID-19 crisis to reposition agriculture. The nation should eschew the merry-go-round oil cycles where there is no movement but just mere motion, to one where fool proof policies translate to concrete steps in implementing this generation's vision for the post-COVID and post-fossil fuel future.

Cost of Oil in Nigeria

Biophysical changes of the fragile ecosystems are taking place concurrently with physical growth of Nigerian economy. Before the COVID-19 pandemic, remarkable transformations have taken place in healthcare, the number of children in school, population having access to electricity and other related developments. To achieve the exact degree of development needed to scale the hurdle of growing from lower middle-income economy to high middle-income economy, the carbon intensity of Nigerian economy must increase in physical terms. Inventory of GHG gases emitted per-sector reported in the First, Second and Third National Communications to UNFCCC explicitly depicts the correlation of carbon emission and Nigeria's GDP growth. High carbon emission accelerated growth in doing so provided the impetus for transformation of the country. In the revision of the National Account in 2014, serious evidence of a growing economy on track to meeting the target of becoming a developed economy by 2020 can be found.[38] This view was further reinforced in the ERGP whereby energy generated from fossil fuels is to propel growth to new heights.

Given hydrocarbon's critical role to government's transformational goal, uncovering its economic benefit is critical to determining the real environmental cost. Establishing oil's real cost is a double imperative owing to its critical role on other national priorities. Not only will it enable determining the true cost of oil; but more importantly, will guide government in fashioning other alternatives to fossil fuels.

Given the focus of this study, we shall not unearth the environmental cost with the degree of intensity it deserves. Surmise, however, that some salient issues will be handled within the tight confines of this work. Moreover, such a study will require specific macroeconomic and microeconomic analysis targeted at various sectors to bring out their GHG inventory before measuring their carbon footprints. A systematic analysis is, therefore, needed to unpack the true cost of oil. This will link each sector to loss and damage from climate change and other socioeconomic debts the economy will have to pay for such pollution—intensive activities. Most of the economic cost undeniably cannot be accurately captured because of the size of the informal sector estimated to be over 36 million small-medium businesses. What we should note, however, is not the size of the informal sector but the shortfall in technical capacity to accurately capture the cost of pollution both within upstream and downstream sectors of the oil industry. The result of this deficit is the large void in data, which in most cases have been relegated to the polluter to fill.

Though Nigeria is determined to see this objectionable practice to the well-deserve grave, institutional bottlenecks have stymied attainment of the goal of government. The cost of oil can be deduced from physical evidence of oil spills and head counts of gas flaring sites without commensurate environmental cost to host communities, national and international ecological footprint, health related underpinnings, changing temperature and alterations of the fragile ecosystem leading to extreme cases of biodiversity loss.

If we are to accurately capture the real cost of oil, then it is necessary Nigeria sketched in the outstanding components of the PIB a new mechanism of determining such pollution, its computation and remedies for such actions while also prescribing stiff penalties for non-compliance within the law. Nigeria's growth-based orientation is closely related with the exhaustion of extra-human nature hence risk assessment must be cognizant of the temporal dimension of pollution while also explicating equity dimension within present generation. By implication, the equity dimension should embrace both present and future generation's needs. The want of publicly accountable pollution burden together with its cost is the source of the intense eco-politics surrounding allocation and distribution of petroleum resources.

This study posits therefore that the real cost of oil in Nigeria is not the cost to present generation but how future generations relate with our utilization of their ecological assets. The environment of their era and the available resources for meeting their basic needs are essential. Answering this question is also with a trans-boundary component for different states and countries in the sub-region. Host communities bear the immediate cost of pollution. For this reason, issues of fairness, equity, climate justice and allocative justice must be considered when finding out the true cost of oil, for instance. Other issues include: who is to bear the cost of environmental protection, the producing country or oil firms or the host communities?

This, again, raises another vital question of whether markets are efficient in undertaking repairs of environment in other to improve the quality. In an economy where the market for environmental services are either rudimentary, non-existent or appropriated or misappropriated for personal stakes determining the cost of oil pollution or oil spills would only be from the lens of the polluter who paradoxically is also in control of government. In such a case, the degree of openness of government in harnessing environmental resources is material to determining the cost of oil.

Besides, pollution is a negative externality- the product of incidental and non-incidental activity of an economic agent (oil firm) which advertently or inadvertently causes or imposes economic cost on another entity (the Niger Delta People). Accordingly, the cost of oil consists of externality related issues. The oil cost analysis resulting from externalizing environmental pollution should include GHG emissions, past, present and future. Though the historic share has no consequences for efficiency even then has serious equity implication. For example, pollution of Ogoniland may not have immediate repercussion in terms of climate change but could in the long-run affect the Northern or Western regions of Nigeria. The cost of carbon should be the impending ecological debt accruing Northern/Western regions in addition to physical alteration of the Ogoniland ecosystems.

State of the Environment in Oil Producing States

Traditional Nigerian societies see nature through Africa's bio-communitarianism world view where human ecological requirement is within biophysical limits while also achieving eco-development through eco-efficiency. A core element of African indigenous tradition and belief system is the importance of living in harmony with nature. The most prominent position on indigenous tradition and belief is the advocacy of returning to the indigenous tradition of "reverence to nature".[39] In pre-colonial Africa, the environment was not an exclusive entity far removed from the people nor was it an object of man's unbiased abuse but one of mutual harmonious coexistence. This deference to nature is rooted in traditional African norms and values where nature plays a prominent role in religious, political and economic processes. An approach borrowed extensively from African mythology where forests are the dwelling places for gods who protect the whole community hence functioned as a moderating force in preserving the biodiversity.[40]

In line with this world view, traditional Nigerian societies see the environment through a deeply spiritual lens. Nigeria's spiritual notion of nature is undergoing rapid cultural shifts lately. No doubt, re-engineering of the role nature plays in the socioeconomic milieu has reinvented the environment, generally. In the race to provide the basic amenities of life, government is tampering with biotic and abiotic components of the ecosystem by way of extracting raw materials, both for export and domestic production. The worrying thing here is that this 'export mono mentality' made the country vulnerable to the vicissitude of the global market. No less importantly, the extractive mentality not only created commodity frontiers but led to a 'deterritorialization of the state'.

In this extractive enclave, state intervention ensures captains of industries control extraction of raw materials.[41] Adverse extraction resulting from rent seeking mentality turned the country into a 'commodity frontier' where irrational enclave logic prevails. In most cases, where this enclave logic prevails there is destitution of communities in the wake of environmental challenges created by these industries. In these de-territorialized enclaves, the local market is inconsequential in the scheme of events with large proportion of materials committed to foreign markets. When companies take over roles within the purview of government, the latter consolidates itself as a police state that represses victims of the system while refusing to meet social and economic obligations. In such clime, even the judicial system ends up enmeshed in the interests and pressures of the private or state-owned extractive enterprises.[42]

This point is worth highlighting. Extraction of natural resources to meet growing needs is as old as the history of man. During the last 500 years, most especially, the colonial period that its all-pervading tentacles denuded Africa of vital resources. To ameliorate the adverse extraction, some states on gaining independence enshrined stiff penalties and regulations guiding exploitation of resources in their constitution. With fossil capital's all-pervading tentacles, implementing these regulations was an uphill task. So, the vital task of this era is how to reduce this adverse exploitation of nature. Exploitation of natural resources will generate underdevelopment in perpetuity and thus is capable of engendering national dead-end. Interestingly, there is some form of 'tropical fatality' in Nigeria's neo-extractive mentality arising from the commodification of nature.

It is a sad development that this neo-extractive mentality makes the country to play a subordinate role in the globalisation of internal and external economies. Despite enormous natural resources, the country is largely underdeveloped due to international market price fluctuations coupled with weak institutions. The challenge here lies in extricating Nigeria from what Bhagwati calls' immiserating growth' where Ricardian rent generated from natural resources such as crude oil will lead to overproduction when prices are higher.[43] The volatility of international markets implies that crisis will lead to producers increasing exports further undermining price to the detriment of other exporters. Interestingly, such sharp practice is of immense benefit to net importers.

Given the above, Nigeria is a victim and agent of change but cannot continue to be both casualty and mediator of change. The country was a victim of change in the past and should not allow itself become a permanent negative change agent. A solution is to curb wanton pollution in the Niger Delta region. Such an action will change the pollution intensive path of crude oil extraction to a more environmentally benign approach anchored on state of art oil technology to deliver the country to the promise ecological nirvana. In changing the carbon footprint, the country should determine the relationship with nature within the context of the respective local cultures of oil producing regions and international best practices in the oil sector. Success of this strategy depends on formulating a coherent approach with strong local anchorage. New institutions and new ways of doing business based on the strategic idea of participating in the global market are keys to taking Nigeria's transition forward. This way, Nigeria's strong eco-sufficient root is not subordinate to prevailing oil greed agenda still making the rounds lately.

Of note, the brief euphoria from the Ricardian rent collected should not propel the government to overvalue rent in place of nature neither should they undervalue human efforts and needs. Since the environment needs mankind, we also need it, however, the nature and form of our transition to market-based economy will determine how nature smiles at us. Decision on extraction and maintaining the environmental quality will not be through any of these two constituencies. Decisions on maintaining a stable environment will be at the political arena where parochial eco-politics has proved to be decisive in influencing extraction hence the outcome of mitigating environmental decadence.

So, the state of the environment in oil producing communities has not really changed since independence. Though the transition from 'brown' resource intensive and unsustainable approach of the past five decades to a greener future is catching on, gas flaring has not been totally eliminated. Nigeria's low carbon drive encapsulated in various development blueprints is with the objective of catalysing growth using the abundant resources. Despite this, some policies of government tend to promote sustained growth against ecological neutrality. For this reason the ecological terrain remains a big battleground between the diverse competing interests who hold differing ecological norms and values. Unfortunately, these battles are in the backyard of rural communities in the oil producing regions but were not consulted. It is pertinent to note that oil producing communities' environment is at the national forefront of embracing the impact of oil extraction. A major outcome of this reinforcement of market drive is accelerated discord between nature and humankind such that commitments are shaped and defined by profitability.

Activities of the oil industry have critically affected a clear majority of the people who are dependent on natural resources for their food and livelihoods. A major source of economic tug-of-war between host communities and oil producers is physical state of the environment in these communities. Judging from dominance of the latter, it may seem government has taken sides with the oil firms as a result cannot effectively navigate between the thin line of maintaining the environment of the South-South region in a pristine state rather is interested in upholding short-term gains of increased oil revenue. This created structural tensions at innumerable scales between government and host communities, host communities versus oil producers, nature and humankind, elite's preference versus national interest.

At another more serious level, there is crisis of balancing growth towards negative emissions in an unequal national policy with varied ramifications for diverse actors and sectors. For the smallholder farmers in these communities, reinforcement of despoliation at the behest of government is a direct threat to their livelihood. In truth, creating an amiable clime for extraction of hydrocarbon will destroy the biodiversity, farmland, fishing ponds, marine aquatic life and more important, the health of the people. Collectively, these problems occur in multiple chains underscoring Niger Delta people's assertion that oil firms are fleecing their region of life. There is merit in this assertion owing to multinational oil firms' propensity of systematically denuding ecosystem of vital life support systems of the region.

The worrying things here is the pace of pillage is growing daily instead of decreasing with the prevalence of environmental maladies. Niger Delta region presents a manifold challenge to policy owing to influence of diverse actors at local, state and national levels that are all jostling to control nature. These groups see the environment as a positive externality in their zero-sum game of exploiting petroleum resources to promote private gains. And where profitability logic subsists, environmental degeneration through large scale appropriation of surplus would degrade the quality of life for the clear majority. But for a select few, it is an avenue to profit from nature. This is half of the story. In real terms, oil is tied to the political fortunes of the government in power in a manner that short-term gains usher in electoral benefits. Here, political fortunes are hijacked from the barrel of oil- the new arsenal in Nigeria's political space. In the place of pursuing long-term ecological wellbeing, oil provides a meaning to politics arming politicians with the necessary tools to maintain their stranglehold on power.

At moments like this the words of Sheikh Zaki Yamani of how he would have preferred Saudi Arabia had discovered water instead of oil rings out clearly to us from the grave. In the face of rising uncertainties in the oil industry one wished that the country had placed more hope on agricultural production instead of oil. Most of the Niger Delta communities who are mainly fisher folks would have wished for the Yamani curse of water instead of oil. Long standing systemic failure resulting from weak institutions turned a seemingly economic positive resource into an ecological negative since externalities resulting from the extraction are not captured through the market. Ensuing environmental degradations from the neo-extractive mentality are diurnal reminders of host communities' vulnerability. Even when and where structures such as OMPADEC, Ministry of Niger Delta, NDDC and other palliatives are instituted and some remediation carried out but graft within the local elite in the oil producing region have stymied a truly people-centric development.

The drama unfolding in the National Assembly which was reported by all the national media of how over N87 billion spent between February and July 2020 by NDDC on frivolous items points to the governance deficits of managing the Niger Delta region. The huge sums of money diverted to personal ends visibly shows the problem is not just the lack of government intervention but more of appeasing the political elite's neo-patrimonial cleavages. Hence the problem is not so much want of viable solutions but is more of institutional failure resulting from the influence of army of rent seekers with assorted motives for our social ecologies. Owing to the neo-patrimonial character and soft state structure, weak institutions have now become tools in the zero-sum game of the plutocrats.

It must be admitted here that corporatist profit drive policy hence the policy space is tainted by the instruments crafted to curtail harmful pollution. To carry this argument further, it seems a direct correlation exists between environmental pollution, shareholders profit, and pecuniary dynamics of the ruling class with corruption as the oil greasing all these into a collective whole. One striking and sometimes disturbing fact is that the deprivation of the Niger Delta people and environment is mutually relational. As more people are increasingly marginalized, other avenues for dislodging perpetrators is hatch creating unending musical chairs where conflicts undermining both nature and humankind are commonplace. For sustainability of ecological resources, government must rebalance various contending forces in ensuring eco-sufficiency through promoting eco-equilibria relations.

Notes

1. The IPCC (2007) concludes there is new and stronger evidence most of the warming observed over at least the past 50 years is attributable to human activities. The Stern Review, (2006) further asserts that IPCC confidence is based

on several decades of active debates and efforts to scrutinize the detail of the evidence and to investigate a broad range of hypotheses. The evidence from IPCC TAR has been collaborated by IPCC AR4 reaffirming and the Fifth Assessment Report further confirms the evidence of anthropogenic induced climate causation in addition to natural forcings.
2. Stiglitz, J. (2006), *Making Globalization Work,* London: Penguin Books Ltd.
3. Most scientific investigations are of the view future temperature increases will depend on the point at which stocks of greenhouse gases stabilizes. Human Development Report (UNDP, 2007) notes that at whatever level, stabilization will require emissions must be reduced to the point at which they are equivalent to the rate CO_2 can be absorbed through natural processes, without damaging the ecological system or the carbon sinks. The longer emissions remain above this level, the higher the point accumulated stocks will stabilize. It was further reported that over the long term, the Earth's natural capacity to remove greenhouse gases without sustaining damage to the ecological systems is probably between 1 and 5 Gt CO_2e. Concluding that with emissions running at around 48 Gt CO_2e, we are currently overloading the Earth's carrying capacity by a factor of between 10 and 50 and that if emissions continue to rise following current trends then stocks will be increasing at 4–5 ppm a year by 2035—almost doubles the current rate. IPCC 2014 further finds that if remedial actions are not taken the world we would be open to climate risk magnifying to unimaginable proportion in the near future causing existential threats.
4. IPCC, 2007.
5. Okoh, A. I. S(2013). *Political Economy of Climate Change in Africa*, Lambert Academic Publishing, Available at: https://www.amazon.com/Political-Economy-Climate-Change-in-Africa/dp/3,659,373,117.
6. Ibid.
7. He Teburoro Tito, President of Kiribati refers to climate change negotiations as contest where the contestants are not mindful of the well-being of mother Earth.
8. Weber, A. (2013). in the article The Economy of Wastefulness: The Biology of the Commons describes the post-Hobbesian dystopia resulting from maximzing of individual benefits leading to the break down of law and order. see https://www.boell.de/de/node/2777222.
9. IPCC-Intergovernmental Panel on Climate Change (2007), *Climate change 2007: the physical science basis*, Cambridge University Press, Cambridge.
10. IPCC- Intergovernmental Panel on Climate Change (2014). Fifth Assessment Synthesi (longer) Report. Retrieved from IPCC: https://www.ipcc.ch/pdf/assessmentreport/ar5/…/SYR_AR5_LONGERREPORT.Pdf.
11. Reports from World Energy Outlook 2015 and 2016 held that the global share of solar is rising every year to tally with increasing demands for renewables. According to WEO 2016, rapid deployment brings lower costs: solar PV is expected to see its average cost cut by a further 40–70% by 2040 and onshore wind by an additional 10–25%. Subsidies per unit of new solar PV in China drop by three-quarters by 2025 and solar projects in India are competitive without

any support well before 2030. Subsidies to renewables are around $150 billion today, some 80% of which are directed to the power sector, 18% to transport and around 1% to heat. With declining costs and an anticipated rise in end-user electricity prices, by the 2030s global subsidies to renewables are on a declining trend from a peak of $240 billion. Renewables also gain ground in providing heat, the largest component of global energy service demand, meeting half of the growth to 2040.
12. IPCC FAR 2014.
13. Donovan (2017, p.1).
14. The Petroleum Industry Bill attempted to reinvent the fledging wheels of NNPC.
15. Heinrich Boll Foundation and Friends of the Earth International. (2015) Coal Atlas: Facts and Figures on a Fossil Fuel Available at: www..Nigeriacoal.org Accessed 12th October, 2017.
16. Pearson (2000).
17. Kale, Y. (2016). *Unlocking Nigeria's Potential...growing through diversification.* A Paper delivered by the Statistician General of the Federal and CEO of National Bureau of Statistics at the 7th Stanbic IBTC Bank Investors Conference, Eko Hotels and Suites Victoria Island, Lagos on the 23rd February 2016.
18. NASPA-CCN (2011) National Adaptation Strategy and Plan of Action on Climate Change for Nigeria. Abuja, Nigeria, 87 pp.
19. Kale, Y. (2016). Unlocking Nigeria's Potential...growing through diversification
20. DFID (Department for International Development), (2009). Impact of Climate Change on Nigeria's Economy.
21. Okoh (2015). Sustainable Development and Adaptation Strategies of the Poor: A Study of 2012 Lower Benue Floods in Nigeria. International Journal of Scholarly Research. Vol.1(1), pp. 1–15.
22. 'Etuonovbe, A.K. (2011),The Devastating Effect of Flooding in Nigeria FIG Working Week 2011 Bridging the Gap between Cultures Marrakech, Morocco, 18–22 May 2011.
23. Ibid.
24. Okoh, (2015).
25. Okoh. A. I. S, Mailumo, D. & Orokpo O. F. (2017). Nigeria's Nationally Determined Contribution (NDC) in the Transition to a Low Carbon Economy. Dutse journal of Economics and Development Studies (DUJEDS) Vol. 3 No. 1. PP. 27–41.
26. Okoh. A. I. S (2017). Makurdi Flood: Disaster waiting to Happen? Available At: https://www.linkedin.com/pulse/makurdi-flood-disaster-waiting-happen-sadiq-austine-igomu-okoh/.
27. The World Energy Outlook Energy: Climate and Change, (2015). A Publication of International Energy Agency Paris, France. Retrieved from https://www.iea.org.
28. Federal Government of Nigeria. (2012). Climate Change and Sustainable Development in Nigeria. National Report for COP18/cmp 8. November, 2012.

29. Bloomberg, 2017.
30. The World Energy Outlook Energy: Climate and Change, (2015).
31. Hansen, (2012a).
32. Federal Government of Nigeria. (2015). *Approved Nigeria's INDC Submision.* Retrieved from www4:unfccc.int/submisions/INDC/submissionpages/submissions.aspx.
33. Ibid.
34. Collier, P. (2008) Oil, Growth and Governance in Nigeria In P. Collier, C.C. Soludo, C. C. and C. Pattillo (eds.), *Economic Policy Options for a Prosperous Nigeria* New York: Palgrave Macmillan.
35. Ibid.
36. Bill Gates, (2018). Speech by Bill Gates, Co-Chair of Bill & Melinda Gates Foundation, at the National Economic Council Aso Rock Presidential Villa, Abuja, Thursday March 22, 2018.
37. Ibid.
38. Dr Yemi Kale, Statistician General of the Federation/Chief Executive Officer, National Bureau of Statistics in a paper titled: Pushing Nigeria's Economic Diversification Forward: Issues & Options at the 7th Stanbic IBTC Bank Investors Conference: held at Eko Hotels and Suites Victoria Island, Lagos 23rd February 2016 sees disconnect between development theory and economic development due to the prevalence of misplaced priorities. He sates: 'The revision of the National Account statistics shows Nigeria's economic structure has changed since 1990 being less dependent on agriculture and more on services. Agriculture used to be about 35% of GDP, it is now 22%. Industry which included crude oil and gas production used to be 36% of GDP but is now 25% while services that used to be 39% of GDP is now over 50%. This however reveals disconnect in development theory. Countries are meant to transform from agriculture to industry then services, and each stage of this transition usually takes some time. However, we seemed to take an alternate path from the agricultural stage and, without really developing the industrial base, gone straight to services, and now we are going back to industry and manufacturing. What happened? The growing demand for services under normal development theory should have been fulfil by manufacturing which should have grown alongside the services. But due to constraints to manufacturing, services supported by imports. So, we want TVs and furniture or cars and, rather than support our local industry to produce to meet this demand, instead we import. Hotels provide accommodation services but most of the furniture in the hotel are imported. So, it provides the service to Nigerians, increasing our services GDP but not manufacturing' (p.8).
39. FGN, (2015).
40. In the work "Some African Reflections on Biomedical and Environmental Ethics," Tangwa's outlook of pre-colonial traditional African societies is encapsulated in what he calls "eco-biocommunitarianism." This metaphysical worldview involves the "recognition and acceptance of the inter-dependence and peaceful coexistence between earth, plants, animals and humans". According to

Tangwa, (2004) this metaphysical outlook underpins cosmic relations between human and his fellow humans. It is also responsible for why traditional Africans are more cautious in their attitude to plants, animal and inanimate things and the various invisible forces of the world.

41. Okoh, (2013) The Political Economy of Climate Change in Africa.
42. Acosta, A. (2011). Extractivism and neoextractivism: two sides of the same curse. Available at: https://www.tni.org/.../beyonddevelopme...
43. Ibid.
44. Bhagwati, J. (1993). The Case for Free Trade. Scientific American 269 (5): 42–49.

Chapter 4
Fossil Fuel Exit: Which Way Nigeria?

> The experience of four decades has shown that exporting oil by itself does not transform poor countries into flourishing economies within a generation. In earlier years, many experts thought the "black gold" of oil would bring riches and economic development. Today their expectations are far more restrained. Oil-exporting countries are more likely to be described as suffering from "the paradox of plenty," "the King Midas problem," or what Juan Pablo Perez Alfonzo, the founder of the Organization of the Petroleum Exporting Countries (OPEC), once called the effects of "the devil's excrement." Their reality is sobering: countries that depend on oil for their livelihood are among the most economically troubled, the most authoritarian, and the most conflict-ridden in the world.
> Terry Lynn Karl, Understanding the Resource Curse

Changing Landscape of Climate Policy in Nigeria

Terry Lynn Karl treatise on *Understanding the Resource Curse* quoted in the opening of this chapter eloquently explicates the dilemma of oil producing countries. Nigeria, an oil producing country is suffering from an acute dose of the 'paradox of the plenty.' Over the years the country has been both villain and victim of its own economic complexities. Why is a country blessed with abundant resources suffering from the 'paradox of the plenty'? Can a country with promising potentials still be one of the most economically troubled and also conflict-ridden states in the world? Where did the country derail? What makes oil a cursed resource in Nigeria and is a source of blessing to others? How can we overcome the 'King Midas problem'?

Nigeria's economic space is changing. As of late, the country is making significant progress in human development only for it to erode. The economic instability in the country is a source of concern to well meaning Nigerians. Dr Kale who is the Statistician General/Director General of National Bureau of Statistics painted a vivid picture of the shortfall in the development policy. In an address to the Stanbic Bank

A. S. Okoh, *Oil Mortality in Post-Fossil Fuel Era Nigeria*,
https://doi.org/10.1007/978-3-030-60785-2_4

management points out flaws in the policy reasoning which he attributed to deficiency for not planning or anticipating 'economic and financial storms' ahead of time. From the statement of one who is the repository of national data, it is evident that Nigeria's petroleum-led economy is instrumental to the impressive GDP growth recorded in 2014; however, it is also detrimental to national development. According to Dr Kale, the policy gap is mainly due to the country's proclivity for governing through crisis. The statistician also noted how the country has not learnt anything from the recent past. And that if the country put the lessons of recent past to good use, it will be clear crude oil revenue which provided additional economic muscle for the GDP growth can also inflict further damages on the economy.

This assertion is not rocket science. What Dr Kale has done is to use readily available data to carry out an analysis in other to arrive at a prognosis. Since government has not put to good use available data on its economic potentials, the nation is actually governing in crises and the mechanism to address these crises is through frequent adjustments in policies. This explains why the nation lost the position to South Africa as Africa's top economy only to recover when oil prices picked up in 2017. In the aftermath of this fall, data available from the World Bank statistics in the mid-year of 2016 rated Nigerian economy the second largest economy next to South Africa with a GDP of $296 billion U.S. dollars. Then, South African economy was the largest economy with a GDP of $301 billion U.S. dollars during the second and third quarters of 2016.

Then, the estimated annual growth rate of Nigerian economy in the first and second quarters of 2016 was −2.06% of the Gross Domestic product (GDP) while South Africa was estimated at 3% growth rate. This is a reversed trend, a testimony to the 'boom' and 'bust' cycles associated with buying and selling of crude oil. With soaring price of crude oil, Nigerian economy at the tail end of 2016 became the largest economy in Africa with a GDP of $415 billion U.S. dollars. Again, the 2020 collapse of oil prices and unrealistic price regime impacted negatively on the economy resulting in a downward trend of petroleum resources.

While agreeing broadly with this, the problem of Nigeria is basically lack of vision to formulate and implement policies with future generations of Nigerians at heart. Lamido Sanusi's (the former Governor of the Central Bank of Nigeria (CBN)) assertion of how the major factors accounting for relative decline of Nigerian economies are political instability, lack of focused and visionary leadership, economic mismanagement and corruption[27] is in line with our findings. Beyond this, we note that even when some policies are formulated, they still lack the necessary foresight to foster endogenous growth with issues of political continuity and policy inconsistency posing serious constraint to economic development. In the process, it is making the policy space to be a graveyard where varied policies are at different stages of decay. It is argued how irresponsible and ill-purposeful leadership are to blame. The basic fact here remains: Nigerian economy is a satellite economy of international monopoly capital. As a satellite economy, it is an economy characterized by primary production highly import-dependent. Nigeria's reliance on oil exports makes the economy vulnerable to the cyclical variations and swings or shocks of the global economic crisis.

With acute policy deficit, present generation are the gatekeepers of their own graveyards. The way and manner we build lasting structures today will determine how history smiles on them. There is, therefore, the imperative to limit our wanton destruction of future generation's wealth. This necessitates holding power to account for the deficiencies of the day. Limiting increases in the global mean temperature is one of the deficits this generation must redeem. A primary concern for redeeming present generation's ecological debt is tendency of exacerbating spread of economic spikes. On this ground, serious urgency is attached to making the economy ecologically friendly.

What this means essentially is, GHG emissions should be within an acceptable limit. The build-up of CO_2 emission in the next thirty years should be in the range of 480 ppm of CO_2. At the end of thirty years, an additional increase to about 550 ppm over the following fifty years is almost unavoidable. It is most likely the global solution to reduce emission of GHG will also work against take-off of development in Nigeria. So, decarbonisation through phasing out petroleum products could be a weighty economic decision to make with huge implications for people and national income, if not properly planned.

With strong business influence, the extractive mentality which is waning in some sector (coal) could re-emerge on the national stage. Partly, this is because the NDC is supposedly based on each country's material possibility to achieve the aims of the UNFCCC. In real terms, Nigeria's submission and subsequent implementation is determined by different factors within the extractive industry. One important determinant is the superficial handling of oil pollution and host communities' challenges in the Petroleum Industry Governance Bill (PIGB) which has not been signed into law by the President. To put this in different context, it would be a folly to assume Nigeria's NDC is a magic bullet to curb all the institutional debilities causing high uptake of carbon. What is noteworthy from this perspective is that the policy space has not really changed to tally with the global trend because limited transparency and accounting mechanisms exist to verify the claims.

In addition to Nigeria's pledge to reduces her GHG emission by 20% (unconditional) and 45% (conditional) in the NDC commitment by 2030, several policies and initiatives already exist which the country can leverage on to achieve the NDC ambition. Such policies include: the National Energy Policy (2003), the National Electric Power Policy (2001), Electric Power Sector Reform Act (2005), the Renewable Energy Master Plan (2005), the draft Rural Electrification Policy and the Renewable Energy Policy Guideline (2006). In 2011, the National Adaptation Strategy and Plan of Action for Climate Change in Nigeria (NASPA-CCN) was approve and in 2012 Nigeria's Climate Change Policy and Strategy was approve by the Federal Government signalling the renewed commitment of the Government towards combating climate change and its impacts in Nigeria.

To overcome waves of adverse dominion of nature, implementers of the national policy should have a clear and undiluted view of the future. With such a compass, the National Renewable Energy Action Plan (NREAP), Vision30:3030 and National Energy Policy including the SE4ALL AA targeted at ensuring universal access to modern energy services; doubling of the global rate of improvement in energy efficiency; and doubling the share of renewable energy in the national energy mix by 2030 will concretely take root. These policies are in line with Energy Policy of 2003 which is with the objective of reducing Nigeria's dependence on fossil fuel thus is set to improve security of energy supply. The REMP sets out short, medium and long terms targets of the national energy supply mix with measures to meet the set targets. According to government's blueprint on SE4ALL AA, the REMP is a roadmap for the actualization of commitment to create enabling environment for sustainable energy supply through deployment of renewable energy for national energy development with active participation of the private sector.

It should be stressed here too that scale-up of the energy investment calls for divesting from energy intensive infrastructure towards a low carbon pathway which is not adding additional adverse carbon load to the overburdened national carbon budget. Given this inherent deficiency, the present clamour to end fossil fuel age will not end the systematic plundering of resources but only places additional burden on the overburdened bureaucracy. Not only would there be manifold economic crises, to some extent series of challenges await government with the end of Oil Age. In a nation that economic transformation is at the take-off stage, the option is quite clear. But building the economy on universally tested modernist's path is prone to mishaps since fossil fuel path is the only route to economic rejuvenation. Taking this argument further, it is an option many developed economies latch on to in the past to attain their present level of development. In developed economies, eliminating fuel fuels from the energy mix is a viable option if it can promote economic growth while also having the co-benefit of curbing carbon emission.

However, resolving crisis involving cutting fossil fuel is entirely a different matter in some developing economies such as Nigeria. Nigeria's post-fossil society cannot take root without first addressing complex factors underlining oil production and consumption. Limiting carbon dependency is, therefore, a thorny issue given the crucial role of transforming the national economy. This is debatable. At a later stage in this book we shall unpack oil's perplexing dilemma when we look at the resource curse complex. But surmise to note it is economically expedient Nigeria maintained the prevailing carbon intensive path despite its consequential impact.

It is, however, an entirely different issue at the global level where the world community face the quandary of embarking on two inversely related roadmaps. In the haste to replace petroleum civilization with a new one, the world leaders seem to have forgotten the history of stone, iron, steam and coal. In resolving this carbon paradox, the world has limited options. One of the solutions proffered at global level is to capture and store carbon emissions. It could be through carbon sequestration programmes facilitating large-scale capture and storage at an industrial scale. The second and most popular option is to reduce uptake of greenhouse gases with the implication of drastically reducing production and consumption of fossil fuel. While the third option is the Reducing Emission from Deforestation and Forest Degradation (REDD+) mechanism aimed at earning carbon credits for maintaining forest resources in its pristine state.

One of Nigeria's primary sources of emission is from crude oil. (Crude oil is a naturally occurring unrefined petroleum product composed of hydrocarbon deposit and other organic materials. Generally, crude oil through refining produces usable products such as gasoline, diesel and various forms of petrochemical). A solution to carbonization of the global environment as stated earlier is the removal of carbon from the global economy. As a result, new waves of decarbonization are in the offing with the outcome of Copenhagen and Paris Agreement reinforcing our drive to curb emissions. At Katowice, Poland (the coal capital of the world) the world leaders gathered to eliminate the source of immense wealth to this country. What a paradox!

Not surprisingly, the conference came to a frustrating and inglorious end. Some argue that the conference is nothing but another merry-go-round jamboree to display sentiments of major power players. Furthermore, USA intransigence armed fossil-led economies with a new weapon to raise their heads from the poisonous fumes emitted while other countries are waiting on the side-line to see how the world reacts to these deviants. No doubt, Katowice outcome is thwarted by oil-producing countries/fossil-based industries' greed and insensitivity to the existential threat. Complacencies of USA, Russia, Kuwait and Saudi Arabia did not help matters resulting in the weak, indefinite and non-binding commitments. Again, the issue of rulebook for monitoring, reporting and evaluating NDCs was weakly resolved thus creating a new loophole for those bent on derailing the global goal. With this development, Africa and the rest of the developing world watched with apprehension the non-binding commitment on climate finance as this will erode political support within their countries. The conclusion reached is that the journey to curb emission agreed upon at Paris is nothing but another dead end lacking any concrete plans for carbon cuts.

Oil Sector Regulatory Regime

Conscious efforts to formulate appropriate laws to guide extraction and exploitation in the oil sector are ongoing. These efforts are outcrop of extensive internal and external pressures exerted by different forces. The history just like the history of oil discovery dates to the pre-independence era when Nigeria's regulations on preserving the natural environment were first formulated. These legislations include: The Forestry Law, 1938, and the Burning Bush Adaptive bye-laws 1958 in Oyo State. These laws had the inherent deficiency of focusing on environmental issues alone thus not particularly designed to impair oil related environmental malfeasance. Even when Nigeria started formulating laws to guide and control oil production at independence, they were heavily slanted towards interests of the oil industry. These laws were designed to favour oil producers detrimental to host communities' preferences. Under conditions of complex interdependence policies formulated in a state will reflect interest of the major constituency exerting serious influence over their leaders.[1]

In the case of Nigeria, the oil industry due to their control of the source of income has an overbearing influence over conflict resolution mechanisms. Control of instruments of coercion is material to laws governing oil production resulting in a distributional crisis. However, distributional, allocative and externalities crises are recipes for armed conflicts with the instruments of governance used at behest of those who dominate the oil sector not by the local communities. Overcoming this crisis demands fundamental reorganization of the way of doing business. For this solution to be effective, a vital class- the host communities who are marginalize must be part of oil management. More often than not these communities pay a heavy price for oil industries expropriations and will continue to pay such prices into the near future. Given the current policy blind sight, oil producers will continue to exploit Nigeria's weakness in circumventing industry standard practice requiring oil firms to deploy up-to-date technologies capable of minimizing serious adverse effects of crude oil extraction. Till date, this best practice requirement has not been complied with rather oil spills are on the increase with carbon leakage as a common trend.

Oil spill first came into national policy space with the legal framework for prevention and remedial action against oil spills with the formulation of the petroleum Act No. 51 of 1969. The Act made it mandatory for oil companies to adapt practicable precaution including provision of up-to-date equipment approved by the Head of Petroleum Inspectorate to prevent pollution of inland water. This provision was largely ignored with gases flared at discretion of the oil industry. As a result, Niger Delta people are traumatized through gas-flares because pollution, noise level, temperature rise, loss of darkness as the whole place is lit up, acid rain, destroyed farmland/crops, and corroded roof and health challenges are adding to the burden of the vastly marginalized population. Roughly 1000 Standard Cubic Feet (SCF) of gas is produce with every barrel of oil. The country produces about 2 million b/d which means 2 billion SCF of associated gas is produce daily and most of the associated gas is flared.

To reduce the gases flared, a broad and more embracing legislation was fleshed out in the 1990s. The new legislation promulgated under the military regime includes the Federal Environment Protection Agency (FEPA) Decree No. 59 of 1992 as amended by Degree No. 14 of 1999 which created agencies to manage all environment related issues. In addition to FEPA, other laws and regulations for control and management of the oil and gas include: the Associated Gas Re-injection Act of 1979, LFN 1990, allocation of Revenue Act No. 1 of 1982 cap 16 LFN 1990, Energy commission of Nigeria Act cap 116, 1990 Hydrocarbon oil refineries Act No. 17 of 1965 cap/170 LFN, 1990, mineral oil (safety) Act, cap 350, LFN, 1990, Deep-off-shore and Inland Basin Production sharing contracts Amendment Decree No. 9 of 1999, and the minerals and mining Decree No. 34 of 1999. Despite all these laws and regulations there were still flaws impairing full implementation of the eradication of gas flaring—controlling and managing oil spills and gas flares.

Petroleum Industry Bill: Peril or Prosperity?

No public instrument received serious public interest like the Petroleum Industry Bill (PIB). Controversies surrounding the PIB are multifaceted because different interest groups wanted to have their say on the direction of the Bill. Prior to this, a menu of factors made passing of the bill a complex challenge including but not limited to: economic, ethnical, regional, geostrategic machination and above all national interest. The PIB due to complex underpinnings took over 8 years to be passed into law. It represents concerted efforts to sanitize a sector engulfed in a miasma of controversies. The Bill unbundled NNPC, provides for the establishment of Federal Ministry of Petroleum Incorporated, Nigeria Petroleum Regulated Commission, Nigeria Petroleum Assets Management Company, Nigeria Petroleum Company and Petroleum Equalization Fund.

It also provides that upon recommendations of the new commission, the Minister of Petroleum Resources can "grant, amend, renew, extend, or revoke any license or lease required for petroleum or production" pursuant of provision of the Act or revoke any other enactment. Of note, the Petroleum Industry Bill has four different components namely: Petroleum Industry Governance Bill (PLGB), Petroleum Industry Administration Bill (PIAB), Petroleum Industry Fiscal Bill (PIFB) and Petroleum Host Community Bill (PHCB). As of now, only the PIGB component has been passed into law.

Despite this remarkable achievement of passing the bill into law, it received mixed reception from diverse sectors of the economy. For instance, a report from the Social Development Integrated Centre (SDIC) sees the PIGB as 'seriously flawed'. SDIC based their assertion on failure of the Bill to address needs of the environment, local communities and pollution. Specifically, the bill does not provide for Health, Safety and Environment (HSE) concerns. There is lack of independence for regulators and a glaring neglect of host communities' interest in the proposed institutions. More importantly, the bill stripped the power of Ministry of Environment on environmental issue vesting such power in the new Petroleum Regulatory Commission. The bill despite some laudable achievements is skew in favour of the oil industry since it does not penalize polluters for their malfeasance.

Implications of the PIGB for Nigeria's Future Without Oil

The brief euphoria over passing of the PIGB has finally ended abruptly. As the dust settles, the chicken has now come home to roast. With the end of celebration people are beginning to read the small prints of the Bill and are raising some vital questions. Questions are asked of the status of pollution of the environment, the Polluter Pay Principle, host communities' neglect, dichotomy between FMoE and NPRC, who should intervene in environmental issues and exalted roles of the Petroleum Minister as chairperson of the NPRC. With these unresolved issues, it downed on the country it is not yet ecological Uhuru because the utopian environmental state aspired for in the oil producing region is still an optical illusion. Not only does the Bill fail to embrace international best practices but is shroud with ambiguity and open-ended handling of key issues.

The Bill marks a watershed in Nigeria's attempt to address decades of environmental despoliation, wanton pollution and poverty associated with oil exploitation in the Niger Delta. The bill was passed into law in 2019 eight years after it was presented to the National Assembly. This is due largely to the fact that passing the controversial Petroleum Bill serves political end in a polity highly dependent on crude oil exports. Not only will it legalise an illegality in permitting gas flaring, it is with weightier consequences in the form of waste of natural resources and pollution of the global environment. Government's foot dragging on signing the Bill owes more to the petroleum sector as the income earner generating most of Nigeria's foreign exchange earnings. Government stated argument is of revenue from the oil sector should drive economic growth and as such short-term benefit of gas flaring outweighs long-term environmental threats. In the following subsection we critically examine implications of the Bill for the post fossil fuel future as follows:

Gas Flaring—Gases are flared in the country despite a High Court judgment in 2005 banning it. The expectation of government is for the PIGB to streamline this issue, setting out all the necessary procedures on gas consumption and production. But in the PIGB gas flaring received miniscule attention. In this, oil firms were handled with a kid glove and parted at back with the injunction to go and sin no more. One would have expected stiff penalties imposed on polluters based on the Polluter Pay Principle (PPP) and should state how it is an off-short of externality thus is a criminal offence punishable by law. Instead the current state of affair legalized an illegality thus giving oil producers a new weapon in their pollution intensive activities to unleash further destructions on the region. Soon, oil operations will ratchet up climate change thereby increasing risk to the overburdened ecosystems. As the nexus of gas flaring with GDP growth is high pollution-intensive output instead of reducing will grow in absolute terms.

Opaque period on ending oil spills and gases flared—The positive attributes of the PIGB notwithstanding, gains in economic sustainability have been erode with the negative emissions. The Bill in present reincarnation succeeded in closing the front door on gas flaring while the back door is ajar for other extraneous interferences. With the expiration of 31st December, 2012 deadline given to oil firms, expectation was high for the new instrument to end or drastically curb GHG emissions. Contrariwise, no deadline exists for ending oil spills rather there is ambiguity and open-ended handling within the Bill. In so doing, the Bill has handed the oil prospectors an open cheque of reducing pollution at their discretion. In this regulatory regime, data transparency and data communication will play crucial roles. Impediments placed on the National Oil Spill Detention and Response Agency (NOS DRA) in the Bill will rather accelerate pollution. With this, the exit from fossil fuel-led growth reinforces debauchery with the constituency who should drive change in this sector also the main causation of our pollution intensive development.

Business-as-Usual Approach—Nigeria's oil sector is a pollution intensive sector. That much we have said on numerous occasions. What we have not stated, however, is pollution drives oil operators' profit. Nigerian oil wealth is infinite in a finite world, however. There is a terminal date of 2070 for exhaustion of petroleum resources or reaching our resource limit. From all indications, Nigeria seems not to be planning for a future without oil. This is premise on the PIB objective of creating an enabling environment for investment into the downstream and upstream sectors to encourage growth. Contrary to current trend at international level of divesting from fossil fuels, the country is promoting and even encouraging pollution exacerbating oil extraction. Key issues of concern such as the PPP, remediation/clean-ups and compensation for host communities were swept under the carpet. In this, Nigeria is hopeful this BAU approach will yield additional profit in the short-run whereas long-run effect will be the responsibility of future generations.

Resurgence of Oil Dependence—The objectives of PIGB *inter alia* are: create conducive business environment for petroleum to operate, enhance exploration and exploitation of petroleum resources thereby optimizing gas surplus. This is with the purpose of establishing a progressive fiscal plan for the PIGB and thus re-orients the petroleum industry towards promoting optimization of profit through minimization of cost. Optimization of cost entails scale-up and streamlining of exploitation to catalyse economic growth. Suggested solution for the environmental crisis remains largely apolitical. The prevailing assumption is for challenges posed pollution in the form of climate change should also be another avenue for expropriation of surplus value. In such an event, Nigeria PIB will give vent to an accumulation crisis.

Nigeria's production system is a subsystem of the biosphere as such production is incapable of growing exponentially in a natural system that is finite, at least without enormous environmental and societal costs many of which may well be irreversible. This explains why in the Niger Delta, there is no limit on exploitation of the environment. When the reproduction of profit faces extraneous crisis, the bottom-line of shareholders is at stake. So, the capacity to accumulate in the face of deliberate pollution is not in doubt but how nature reacts to its violation is the million-naira question. We witness replication of accumulation crisis in the Niger Delta where it led to an over-production hence an externalities crisis. Yet, Nigeria's ecological crisis creates environmental movements with the tendency of deepening into economic crisis.

Piecemeal legislation on the PIB is part of the problem as it can create new avenues for further distortions in the oil sector. When matured economies are exiting fossil infrastructure by laying out of plans to curtail fossil fuel production, the country is just formulating a new law to reinvent carbon-led growth. In other countries such as Saudi Arabia they are building new structures in the process laying out their low carbon and growth plan. Nigeria on their part is finalising plans to build more fossil infrastructures with the obvious risk of being stranded assets.

In a country where oil related conflicts are on the rise, solution to the plethora of problems beleaguering Nigeria is not to reinvent fledging wheels of the oil sector but re-orienting carbon-driven growth towards a more productive path with nature. In as much as the country sees intensification of hydrocarbon led growth as pathway to economic emancipation, the economy will contract. With this the much-desired environmental salvation becomes a proposition of doubtful potential because the country would record near-term gains but in the long-term will promote environmental crisis.

That said, political ends must speak to economic means. Where the interrelatedness is subordinate to extraneous logic, scarcity rent will promote pollution through irreversible saturations and biodiversity loss leading to the reaching of our biospheric limit- the oil mortality of Nigeria thesis. In this rationalization, increasing rate of overproduction crisis is part of the BAU 'boom' and 'bust' cycles of capital. There is, therefore, no sustainable capitalism in Nigeria. For sustainable capitalism depletion must ensure sustainable utilization of ecological assets in a manner blueprints such as PIB are not just ecologically accountable but must be people-centric, eco-efficient and eco-friendly. Above all, it must be cognizant of the intertemporal dimension of resource use in the transition to a net zero civilization.

Nigeria's Energy Status

Nigeria is the 8th largest producer of petroleum with oil reserves estimated at about 36 billion barrels. The nation has the 6th largest deposits of gas with natural gas reserves estimated at a minimum of 100 trillion cubic feet. There are over 34 discovered solid minerals, including significant uranium deposits; abundant arable land with over 44 exportable commodities. Large quantities of fossil fuels and renewable energy resources (Table 4.1) are available throughout the country with almost all the regions having one form of natural resources in abundance.

Despite vast resources, the country still experiences acute energy poverty with over 60% of the population either lacking access to modern energy sources or having to cope with inadequate and poor-quality supply. A total of 120 million people rely on traditional wood stoves for cooking as majority of Nigerians depend on fuel wood as source of energy which also have huge implication on deforestation, and many urban households use generators for their electricity needs. In Nigeria, more than half of the population (72%) depend of fuel wood to meet their energy need. The traditional energy source account for roughly 55% of Nigeria energy requirements with the

Table 4.1 Total primary energy consumption in Nigeria (2011)

Energy form	Consumption (TOE)	(%)
Hydropower	5,92,338.50	0.71
Fuelwood	5,45,39,027.42	65.50
Petroleum Products	94,67,060.83	11.37
Coal	4,833.79	0.01
Natural Gas	1,86,63,720.08	22.41
Total	8,32,66,980.62	100.00

Source Bala (2014)

consumption estimated at around 80 million cubic meters.[2] Presently, only 40% of Nigerians have access to the electricity grid. Lack of access to electricity directly affects livelihoods lowering the quality of life. Rural population are extensively reliant on biomass for domestic purposes with the sector, providing 65.50% of the total energy consumption in the form of traditional biomass. Natural gas accounts for 22.41% of the total primary energy consumption (Table 3.2) with it as the source of 64.97% of the commercial energy consumption. Whilst hydropower produces 0.71% yet accounts for 2.06% (Fig. 4.1) of the commercial energy consumption even though with vast potential is still largely untapped.

The huge fossil deposits have immensely contributed to the transformation recorded (Table 4.2) in the country.

Existing grid-based generating capacity of about 8,017 megawatts (MW) cannot bring about the transition to high middle-income economy. To realize the Vision

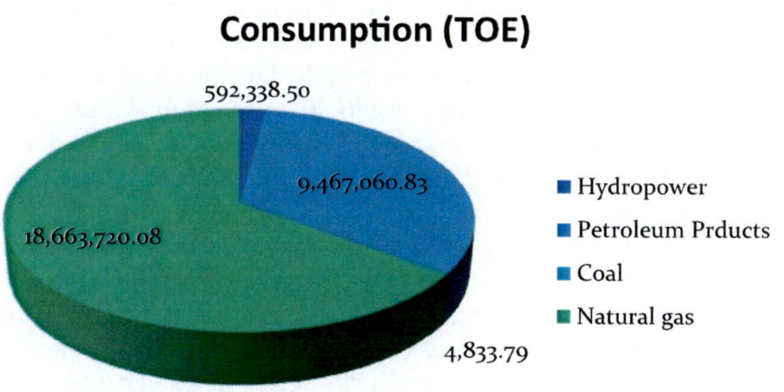

Fig. 4.1 Total commercial energy consumption. *Source* Energy Commission of Nigeria

Table 4.2 Fossil energy resources and nuclear energy sources

S/NO	Resources	Reserves	Production (2011)	Domestic utilization (2011)
1	Crude Oil	37.2 billion barrels	0.721 billion barrels	0.097 billion barrels
2	Natural Gas	187 Tscf	3.100 Tscf	83.0% -Fuel, Industries, re-injection etc. 17.0% -gas flare
3	Coal	2.7 billion tonnes	0	Negligible
4	Tar Sands	31 billion barrels of oil equivalent	0	0.224 million tonnes
5	Nuclear	Yet to be quantified	0	30 kW experimental nuclear reactor

Source Bala (2013)

20:2020 goal, generating capacity needed increase to 20,000 MW in 2020 and 35,000 MW by 2050.

For now, about 78% of the generating capacity is fossil power (fuel oil, gas, and coal) while the remainder is hydropower based. Besides increasing power capacity, this source of power also contributes to the adverse carbon load. Unfortunately, government is considering purchasing 'Tokumbo' (used) coal power plants to generate 30% electricity. Government intends to achieve a 2% per annum energy efficiency ultimately achieving 30% efficiency by 2030 through increased investments in efficient gas power stations.

In the energy sector the proven crude oil reserve is 37.2 billion. The world's sixth largest gas reserve, with 187 trillion cubic feet (TCF) of high-quality, proven reserves is in Nigeria. Of which around half is Associated Gas (AG). The country's natural gas reserves in the offshore and onshore with potentials in the inland basins of Sokoto, Bida, Chad, Benue trough, Anambra etc. Natural gas from these areas utilized for power generation and other industrial uses. About 50,000 MW power generation driven by natural gas will be make available by 2020 but only 8,017 MW are in place. For now, most of the AG is flare while initiatives implemented to curtail flaring have recently started to produce results. To achieve growth targets set within the Vision 20: 2020, gas utilization through increased industrial and domestic use is to increase. Effectively, meeting the Paris obligation in concrete term could be a direct threat to the major source of income.

Another source of energy used for industrial and domestic purposes is coal. Coal was first discovered in Nigeria in 1909.[3] Nigeria's coal reserves are estimated to be more than 2.5 billion tonnes. The resource played an important role of providing electricity in addition to other uses, such as fuel for steel and cement production, and other industrial activities. There are coal reserves in Owukpa, Enugu, Afikpo, Okaba and Okobo and other areas. Production of coal commenced in 1916 from an open cast mine at Obwetti. Sambo opines that at projected production levels proven coal reserves estimated to last 147 years.[4] In contrast, proven oil and gas reserves are equivalent to around 41 and 63 years respectively at current production levels. There is further imperative for plan for the period when oil stock would be exhausted.

The vast coal deposits have been untapped with most existing mines closed and this is due to lack of economic viability. Given this, coal plays a secondary role in the energy generation objective of the country. The National Energy Policy and National Energy Masterplan both promote coal power generation as substitute to gas powered plants. In the National Energy Masterplan, government intends to pursue a comprehensive programme of resuscitating the coal industry. Resuscitation through the private sector is link indigenous participation in the coal industry but it should be carried out in an environmentally acceptable manner. Subsequent plans of action all amplify coal's importance in the energy mix. To maximally utilize coal deposits, government signed a memorandum of understanding with the Chinese firm HTG-Pacific Energy to exploit coal in Enugu. Up till now, no environmental impact assessment to ascertain effects on host communities has been carried out.

Though Nigeria is Africa's most populous country, it faces phenomenal complexities. One of such challenges is energy related. Energy poverty is very high (or simply put, a country in energy crisis) contributing to most of the challenges experienced today. Although vast renewable energy potentials in the form of wind, biomass, solar and hydropower exist with large swathes of arable land (Map 1) exist it has not been maximally utilized. Agriculture is one of the major sectors most vulnerable to the impacts of climate change and it is the sector which also employs the largest labour force accounting for close to 40% to the Gross Domestic Product while over 70% of the working population are engaged in agriculture-related activities. Unfortunately, the contribution of agriculture to the GDP is to decline in the coming years. Agricultural productivity per unit of land, labour, and water need increase considerably to feed a rapidly expanding population. The country's potential to leapfrog into a greener future will be more realisable if they can tap the vast renewable potentials.

Government exploitation of renewables tends to give a different impression of the decarbonisation initiative. It seems government has not given high preference to renewable energies as it deserves. Some may argue that government does not really want the transition to a low carbon economy to happen since renewable energy is not the central plank of the energy reform agenda.[5] To a certain extent; government's focus is to provide energy to fuel growth with job creation and poverty reduction irrespective of the source of energy. Though renewable energy is catching on, the energy trajectory is not moving toward a cleaner pathway at the speed to reduce climate change. An examination of the various sectors will suffice.

In the agricultural space, there are 11 million hectares of woodland forest with enormous biomass potentials. Additionally, there is over 28.2 m hectares of arable land for agricultural produce, which also generates 30 m tonnes/year of Municipal Solid Waste. Traditional biomass is the primary sources of energy, hence the energy sector vital for green growth is evolving. It provides 65.50% of the primary energy supply with over 70% of the population engaged in biomass related economic activities (Table 4.3). In recognition of the immense contribution to national development and the reduction of environmental threats, renewable energy systems are in place in most cities. The objective is to reduce overwhelming reliance on gas to generate power as well as curb gas flaring, which has made Nigeria the second top gas flaring nation globally.

To achieve this objective, a biofuel policy was formulated with the Nigerian National Petroleum Corporation (NNPC) mandated to spearhead the drive with launching of the Nigerian National Biofuels Programme in August 2005. Despite this, biofuel power generation is minimal with power generation of 5 MW from rice husk in Ebonyi state while other biogas projects for heating and power generation are ongoing. More, biogas digester technology has been domesticated and a few pilot biogas plants have been built. Examples are: the human waste biogas plant at Zaria prison, cow dung-based biogas plants at the Fodder farm on the National Animal Production Research Institute (NAPRI), Mayflower Secondary School, Ikenne, Ogun state, an 18m^3 capacity pig waste biogas plant at Ojokoro/Ifelodun Cooperative Agricultural Multipurpose Society in Lagos.

The hydropower potential is about 15,000 MW but only about 3000 MW is currently exploited. A private company, the Nigerian Electricity Supply Company (NESCO) and government have installed eight (8) small hydropower stations with the

Table 4.3 Renewable energy resources

S/No	Resource		Reserve	Utilization level
1	Large hydro power		11,250 MW	1,900 MW
2	Small hydro power		3,500 MW	64.2 MW
3	Solar Energy		4.0kWh/m2/day 6.5kWh/m2/day	15 MW solar PV stand-alone No solar thermal electricity
4	Wind		2-4 m/s at 10 m height	2 × 2.5 KW electricity generator; 10 MW wind farm in Katsina
5	Biomass	Fuel wood		3.4 million tonnes of firewood/yr
		Municipal waste		8.5% cultivated
		Animal waste		
		Energy crops and agric waste		

Source Bala (2013)

capacity of 37.0 MW around Jos, Plateau State. Similarly, a 2 MW station at Kwall falls in river Kaduna and an 8 MW at Kurra falls generates electricity for consumers around the region. A large unexploited prospective needed to be developed fully. For example, the Zungeru, the Mambila, the Gurara II, the Dadin Kowa dam, the Tiga Dam, the Kasimbila and many other small hydro power sites in addition to cascading existing sites. From the hydropower sector, about 14,000 MW of hydro is expected when the Mambila plateau Dam awarded to a Chinese firm is completed by 2020.

Significantly, then, solar energy is to contribute enormous amount of low carbon energy. Solar capacity of the country is in the region of 4–6.5 kWh/m2/day and can be converted to energy for homes through solar PV and solar thermal process. It may be transformed directly into heat using solar collectors or directly to electricity using solar PV cells. The expectation is about 13,000 MW of off-grid PV solar electricity is put in place in 2030. Only about 15 MW of dispersed solar PV system is on the ground, however. If Nigeria is to cover only a small%age of the landmass with solar panels, it could produce 192,000 MW of power, compared to the 8000 MW currently available on the national grid from all sources of energy.

To realize an excellent condition for electricity production from renewable energy sources, national and international investors are already knocking at government's door. Lately, some parts of government have woken to the untapped energy potentials. The upsurge in interest is fallout of the discovery of huge natural gas reserves and renewables. Besides, the Federal Government of Nigeria has declared a gas-to-power revolution and the state governments have toed this line showing serious interest in large-scale solar energy plants. There is 80 KW solar PV power plant in Kastina State University, Katsina from the Japanese Grant in Aid.

Other solar developments include: Circa- 8 MW, Solar PV Street Lighting Solar project, 24kw solar PV mini-grid in Bisanti, Niger State, street lighting projects with capacity up to 10 MW across the country, Blue Camel 40 k rooftop solar system, Pilot mini-grid rural lighting projects, 17.5 kW OGD Farm settlement in Idekan, Ogun State, Green Village Electricity i24Wp solar PV system in the Egbeke community in Rivers State, 17.5 W Esham Community in Cross River State. Though, poorly executed- street lighting in Benue, Lagos, Nassarawa, Sokoto, Abuja, Plateau and a host of other states act as a disincentive for further investments in solar power.

In terms of wind power, Nigeria's wind resource of 2-4 m/s at 10 m height is low for a wind farm electricity generation using conventional wind generation. At higher heights greater wind speeds may be encounter where reasonable electricity could be generated. It is envisioned that 40 MW of wind electricity should have been provided by this year (2020) but only a 20 MW wind farm is in place in Katsina state.

Low Carbon Transition: Myth or Reality?

Nigeria's low-carbon future is sustainable to the level of how political office holders implement the different blueprints for growing the economy on an eco-efficient trajectory. The underlying logic of decarbonisation in some quarters, as it has been argued, is driven by need to meet long-term gains with significant near-term anchorage. The bane of the matter is power generation and distribution. What is more, the era of giving money some form of fairy dust aura when it is sprinkle over the power generation imbroglio there the hope it will magically resolve the looming crisis is now outdated. To be clear, Nigeria's quest to change the power generation and distribution crisis on the back of obsolete machinery is delusional. Also, to roll back years of gas flaring will task commitment to Vision 20:2020 and the ERGP of government. In the immediacy, gas flaring remains the poster child of all that is wrong with Nigeria's GHG abatement policy. For others, the level of actions targeted at the transition to a low carbon growth recorded amounts to 'bailing out a few buckets of water as the Titanic sinks'.

A major constraint is institutional inaction. Institutional inertia experienced in the country cannot usher in a new economy with the ethics of sustainability neither will it promote spontaneous green uprising. Contrariwise, it will push many below the poverty trap with consequential ramifications for natural resources. So, we can say that Nigeria is on the cusp of breaking from neo-extractive tendency towards a new dawn of material consumption wherein ecological sustainability is harmoniously integrate with economic growth. What we are witnessing today is the throes of death and birth pangs. The old era is fighting back by deploying all arsenals at their disposal (including the very instrument of the Paris Agreement rulebook) while the new age is angling to unseat the dying age.

Again, the country lacks clear cut policy and conceptual clarity on how they can transition to the new era. It seems the new age has committed a fatal error of imposing a new paradigm on an old concept without completely exorcising the ghost of the past. What has emerged is that the new age has taken a pick axe to the root of the old age without firmly laying out any foundation of the new era. Overall, one would expect the old and new epochs having some form of harmony with a level playing ground for their socialization of nature.

So, it is fair to say decarbonisation will proceed at different levels for diverse societies. In a country where vicious cycles of capital production persist, pollution will inhibit any real possibility of progress. Arising from this, long-term risks associated with climate change will severely impact development efforts by reordering priorities towards growth and industrialisation.[55] Allied to this is the fact that reduction of gases spewed from oil exploration should be handled as climate security challenge with unambiguous transparency regime outlined to checkmate excesses of the oil firms. Alas, it will lead to decreasing revenue from oil export which is the mainstay of the Nigerian economy.

It will further mean reducing biomass extraction and consumption, which is the primary energy source of over 80% of the rural communities. Finding an appropriate energy mix to navigate the climate stalemate at international arena while mitigating economic growth needs at national level should be the ultimate emission stabilisation goal of Nigeria. Such policy should synergistically integrate the climate Bill into development activities if it must achieve the desired goal. At present, the Petroleum Industry Bill and Nigerian National Climate Change Policy seem to be working at cross purposes. That much needed to be said. Even then, there is still the problem of making sure these bills do not stifle growth and this should inform complementary policies formulated on the abatement of the scourge.

All hope is not lost in the transition to an ecologically sustainable polity. There are loud voices from within the civil society organizations clamouring for an end of the wanton destruction of the eco-system. Within government circles voices of eco-centric functionaries are heard with fight for ending despoliation of resources now becoming a national issue. These visionaries' voices are gradually gaining recognition with sketching of the NDC. As of late, incremental steps to trim down the strong carbon footprint of the nation are made. Whether this influence will not be upturn by oil industry kingpins who are very astute in railroading international obligations like the NDC into other pressing national priorities only time will tell. But time is not on the side of the planet neither is it on Nigeria's side.

For good or bad, the post fossil future is taking shape. On impulse, the solutions proffered for mitigating climate change at the global level have left an indelible imprint on the national decarbonisation initiative. So, what can be considered a low carbon growth is gradually taking shape. Most of these activities are, however, scattered in different sectoral policies without a concretely articulated national roadmap of greening the economy. If acts of sectoral greening can be adjudged a low carbon growth, then a form of green economy exists alongside with the 'brown' economy.[55] Finding ways to protect the global ecosystem; curtail climate risk; improve energy security simultaneously improving livelihood assets of the poor are significant issues in achieving an equitable well-being for the people.

With the urgency attached to climate change, a compromise on pursuing environmental protection against economic growth is being reached. Consequence of this compromise is major decarbonisation efforts such as the Kyoto Protocol is view as a symbolic measure to end GHG emissions. Other major steps have been taken in line with the global decarbonisation initiatives, however. For instance, as a non-Annex I Party to the Kyoto Protocol Nigeria carried out several climate abatement measures (see Appendix 1 and 2). Some of these commitments to the UNFCCC are political commitments while others created some awareness of the negative effects of pollution. Most are political obligations comprising of convoluted system of arbitrary measures meant to create the ambience of climate action while in real term it promotes climate inaction.

Nowhere is this more obvious than in the NDC where spurious commitment to pursue pro-ecological measures is made. The NDC is supposedly based on the material possibility to achieve the aims of the UNFCCC. Nigeria hurriedly acceded to embarking on a deep decarbonisation pathway within the economic and political circumstances. Diverse forces within the nation are the real determinants of the submission. With the influence of interest groups, the NDC is treading on a shaky path capable of affecting its implementation. Within the NDC, several issues constitute constraints to the global initiative, however. The marginalisation of a clear majority of the people who are poor is one such constraint. If nothing else, the Agreement entered at Paris will have various ramifications for different societies at varied stages of growth. In addition, the outcome of COP21 might engender growth in some communities, yet, can create or entrench inequality in others. Regrettably, this crisis is magnifying on the premise adjusting the national carbon budget could seriously impair development. The implementation will have far reaching consequences most especially within the rural poor where the proposal to put an end to fuelwood consumption will criminalise the poor. Such policy places a new economic burden on the people which invariably further entrenches poverty.

Often, decarbonisation measures involve several issues not least is the marginalisation of a clear majority of the people who depend on natural resources for their livelihoods. To continue this warring path with the people will mean the promise of reaching an ecological nirvana espoused at Paris will amount to nothing but another optical illusion. Not only will the submission at Paris be a great burden for the economic wellbeing of the poor, but more critically it is with developmental imperative capable of negating the very essence of the conference. Another conundrum is tackling of the political tug-of-war between economic growth and green growth. Judging from the facts on the ground, the country cannot grapple with uncertainty arising from navigating the thin line between economic growth and degrowth (materialization and de-materialization) while also protecting long-term decarbonisation targets. For sure, there are grey areas in the submission since politics rather than ecological imperative is the major determinant of the national commitment hence the tackling of the emission burden.

Another encumbrance to attaining a low carbon growth is the heterogeneous nature of the economy. The heterogeneous nature of the economy made Nigeria susceptible to shocks. Other challenges are energy poverty, the price of renewable, food and biofuel complex, land use change, investors' apathy, political manoeuvring from an unchanging oil industry, conflict resources, technological handicap and the intermittency shortfall. Collectively, these problems occur in multiple chains underpinning the characterization of the ongoing decarbonisation initiative as a tricky venture. These impediments could result in structural tensions at different levels in most communities. On one hand, it could lead to the problem of how to redistribute emission budget to address consumption and production of carbon in diverse regions at varied stages of growth. While at another- more serious- level is the crisis of rebalancing growth towards net negative emission in an unequal national polity with varied ramifications for different actors and sectors. For the smallholder farmers who rely extensively on biomass, a low carbon growth could have dire consequences for their livelihoods. But to the superrich, it is another avenue for profiting from nature. Although the Paris Agreement has the trapping of grassroots participation, it lacks local anchorage. Since political support for limiting emission at local communities is low or non-existent, collective aggregation of Nigeria's commitments will be lower than the 20% unconditional and 45% condition political commitments.

Building a Post-Fossil Fuel Future

The implosion of oil predicted by Zaki Yamani is now a global reality. It is not just a global reality but it is equally a Nigerian reality. The imminent collapse of the global energy economy is hanging like a 'Sword of Damocles' over the present and future generations' heads.[1] As of late, large swathes of economies are singing different tunes on the outcome of this looming danger. Countries who before now were avid advocates of fossil fuel who were queuing up at oil producers' ports are today eulogising end of oil. But this technological shift is not without its main drivers. Diverse issues drive move to curb fossil fuels amongst which are carbonizations, economic instability within the international oil market and the despoliation of natural resources by resources endowed countries. Little wonder that Juan Perez Pablo Alfonso who was the co-founder of OPEC calls crude oil the 'devil's excrement'. Others see oil as the 'black gold' afflicted with the 'Dutch Disease'- a testament to oil's volatility and unpredictability.[2] Some have argued that fossil fuels would not only bring waste, corruption, degradation but death to host communities.

Despite this negative image, oil is the new elixir equally afflicted with the 'resource curse' which is exposing these economies to the vagaries of climate change. Economic benefits of oil to a producing nation as Kantai asserts, outweighs other benefits. In his words: "With the cost of living already high, and high-level corruption scandals making a mockery of the government's calls for belt-tightening measures, cheap petrol is perhaps the only benefit citizens of the oil-rich federation enjoy".[6] Access to cheap fuel creates a conundrum for policy-makers in Nigeria as it needs oil for growth even then is with harmful effect to the wellbeing of the planet. For another thing, corruption in the oil sector is rampant.[7] "With a near total absence of government presence in the lives of citizens, corruption is the only system that seems to work. Hardly any institution in the country is not weaken by this virus".[8] More the fact, corruption plays a large role in the resource curse since government respond to the windfall of oil by spending on infrastructural development programs with little contributions to the wellbeing of the people. This excess capital propelled rent-seekers into taking advantage of the natural resource money.

Within the academic circle the resource curse received significant attention. One of the key discussants of this issue nationally is Prof. Kayode Shoremekun who in the paper titled: "Nigeria, Oil and the Yamani Syndrome'[9] extensively delved into the resource curse syndrome. In the paper the scholar opined that Nigeria "like a number of oil rich countries have become victims to the "resource curse". In his submission the scholar seems convinced Sheik Zaki Yamani's prophetic declaration on oil as a cursed resource is also afflicting Nigeria and thus is orchestrating the worst aspects of human existence for the poor.And that rather than being s source of blessing oil is a cursed resource replicating conditions of underdevelopment. In his opinion, Nigeria's crude oil producing nation status made the country to have bloated ambition and aspirations leading to misguided priorities. If properly utilized oil should have been the basis of prosperity but has now become one of the main sources of instability.[10]

Oil's intrinsic deficiency is the consequence of the country's inability to use the resource to promote endogenous growth; this made it difficult for Nigeria to be view as an oil producing nation. Arguably, Nigeria's authentic status is one of recipient of relatively vast and providential oil revenue. The operative words here are "relatively" and "providential". And this may well be the reason the country is still in the throes of underdevelopment. In simple terms, providential revenues cannot be the basis of development (See Appendix 3). Bill Gates in his paper delivered to the Federal Government of Nigeria bemoaned the country's non-inclusive growth and overt reliance on infrastructural development which is neglectful of human capital development. According to Bill Gates, "development is a conscious choice and process. The illusion of an oil producing status is also accompanied by other illusions, which form an integral part of our national life."[11] In this assertion, Gates presciently puts a nail to Nigeria's coffin as an oil producing State. Clearly, then, Nigerian economic fortune sits at the crossroad of economic indecisiveness and ecological uncertainties. Even so, it is premature to consign the country to the backwoods of history owing to resilient nature of the people.

Resource Curse: Myth or Reality?

A question most public policy analysts on resource curse ask is why are oil-rich countries worse off than their resource poor neighbours? Could the resource curse be a myth rather than reality? Before answering the meta-questions, we ask: what is a resource curse? The resource curse refers to the inverse relationship between high natural resource dependence and economic growth rates. This is the answer to aforementioned questions provided by Terry Lynn Karl in the book title: *'The Paradox of Plenty: Oil Booms and Petro-states'*.[12] In answering the hypothetical question the author notes how booms themselves cannot be held responsible for failure of the leaders to change the discourse since "increase in the supply of monetary metals does not, any more than autonomous increases in the quantity of any other kind of money, produce any economically determined effects. It is obvious these will be entirely contingent upon use to which the new quantities are applied. And that, revenue from natural resources affect state structures as such it is not only "policy environments" but also "autonomy of goal formation, the types of public institutions adopted, prospects for building other extractive capabilities and the locus of authority" affected by these revenues.[13]

Nigeria's fossil fuel industry is at the mercy of implementers of national policy who hold dissimilar ecological norms and goals for exploitation of this resource. Their divergent positions can derail other goals of government given there are diverse forces aligned to change nature. These forces intent on changing nature include businesses, politicians, civil societies and academics. A complex transformation involving these diverse change agents requires innovative solutions. However, incentive for resolving such complexity instead of stimulating innovation rather will promote disincentive. With its neo-patrimonial character, tackling the resource curse debacle is a multifaceted issue since the national cake is a zero-sum game where the elite is bent on accumulation of both wealth and power. With the country's rentier nature, revenue from petroleum resource will be diverted to private ends not to other extractive capacities of the country. New policies on extraction will not only provide new avenues for businesses to reap more profit but will orchestrate depletion of the environment to new heights. Worse still, such change is a difficult issue since it can stifle structures put in place for its operationalization. Before the discovery of oil in Ghana, the country was considered one of the most stable, corruption free and rising star of Africa's renaissance. Then, the country's economy was growing at stable rate but the discovery of oil drastically changed perception of the country with massive inequalities and corruption commonplace. Similar trend seems to be emerging in Kenya that just of recent commenced exploration and exportation of their oil.

Correlation between resource curse and the Dutch Disease is because resource cursed countries are prone to an acute case of the Dutch Disease. Simple logic behind the Dutch Disease theories according to Terry Lynn Karl is a situation when an economy in full employment equilibrium, a permanent increase in the inflow of external funds results in a change in relative prices in favour of non-traded goods (services and construction) and against non-oil traded goods (manufacturing and agriculture), leading to the crowding out of non-oil tradeable by non-tradeables.[14] An appreciation of exchange rate leads to a decline in competitiveness hence production and employment in the traded-goods sector. The mechanism through which this change takes place follows directly from the model's assumptions of full employment equilibrium and static technology. With these assumptions, external funds (from an oil boom) can be translated into real domestic expenditure only if the flow of imports increases. However, since non-traded goods cannot be imported easily (or only at prohibitive costs), a relative contraction of the traded-goods sector is inevitable, otherwise the resources needed to enhance the growth of the non-traded sector would not be available. Thus, the model predicts de-industrialisation is the inevitable structural change that occurs because of oil booms.[15]

A research conducted by Jeffrey D. Sachs and Andrew M. Warner delved extensively into the correlation of growth with resource abundance. They used ninety-seven countries studied over nineteen year's periods to reach their findings. The duo applied regression analysis in measuring the impact of mineral and other resource export on GDP growth. The conclusion reached is a high "ratio of natural resource exports to GDP in 1971 had abnormally slow growth rates between 1971 and 1989" .[16] In a resource cursed polity there is high incidence of corruption and undue dependence of crude oil. In some of the studied states revenue accruing government instead of fuelling growth is diverted into private bank accounts of those who control instruments of governance. These states are not only neo–patrimonial in character but are rentier states where enclave logic prevails. Weak institutions within these states drive them towards dependence on oil since their public institutions lack capacity to manage an extremely volatile commodity. The result of such lack of capacity is instrument of governance are subverted and controlled by the very business entities it is meant to protect their citizens from.

In a situation economic influence and political power are in the hands of a select few, the lines between public and private entities are blurred. And where the line is blurred, rent seeking as a strategy for wealth creation is rampant. So, to some extent it could be argued that Nigeria suffers from the resource curse.

Professor of Economics Pat Utomi, joined the conversation on diversification with a varied stance on the resource curse debacle. The erudite scholar compared Nigeria and Indonesia who according to him are development twins with different levels of development. He states: "if you look at countries that have managed their affairs very well, you will realise oil managed to be a blessing in some places.[17] Classic example of where it is a blessing is Norway. When carrying out a comparison in developing countries there is discernible difference between Indonesia and Nigeria. These two countries are with similar history, though are at the polar extreme of the economic divide. He further stated "Indonesia has always been considered Nigerian development twin. Both are growing in dissimilar directions. Again, they are both populous and oil gifted countries. Indonesia developed its oil, and it also successfully diversified the economy. Their oil sector is striving, their industry sector is not doing badly and their agriculture is also doing fine.

Oil is not supposed to be a curse, but the way Nigerians have run it has made it, unfortunately, a source of damage rather than advantage". 'My personal prayer', he added, 'is for oil to drop to as low as $4 per barrel and stay there for four years so that we can begin to have an economy. There is something wrong with the way we behave. We are waiting for oil price to go up and we will be in a more dangerous situation". This observation is happening with the novel COVID-19 pandemic forcefully impose economic meltdown. The country is only few months into the global lockdown yet, is recording severe economic shocks. Not surprisingly, the country went to IMF for loan. On the 29th April, 2020, IMF approved additional $3.4b emergency loan to fund Nigerian budget deficit. In this, Professor Utomi's admonished: "when you are in a commodity-dependent economy and the commodity is in boom, you save significant part of your earnings. And you use your savings to attract other savings from abroad and make investment in other areas of growth"[18] was not heeded.

Contrary to this, Nigerian politicians were on a buying prey jetting from New York to Dubai for leisure and health tourism. But they neglected to develop the health facilities in the country. When Coronavirus came knocking, it knew not rich or poor neither was it a respecter of person or national boundary. With dilapidated state of medical facilities, the health institution which epitomises national deficiency that crumbled at the first sign of stress. With the political class finding no option of where to escape, the country's medical centres became a much sought-after institution and even desirable.

Gavin Wright and Jesse Czelusta criticize the resource-curse literature on rounds it paid little attention to economic character of mineral resources or to the concept of "resource abundance."Theirs is a black box approach. Virtually without exception, this study equates export of mineral products with "resource abundance," seen as a simple reflection of an exogenously given geological "endowment." When the revenues from this activity are describe, terms such as "windfalls" and "booms" are generally not far behind.[19]

This synonymy is a matter of implicit assumption rather than analysis or demonstration, generally unquestioned, and all too often unrecognized. On closer scrutiny, each step in this chain of equivalences is questionable. History is replete with instances of crises/doom turned into breakthrough/boon engendering a radical revolution of society. We cannot vitiate a resource for being source of 'boom' or a 'boon' if we do not examine the root cause of failure in one society and success in another. A resource on its own is not doom per se, but a society could be doom because of visionless leaders. Ascribing sundry rationalization to a commodity is to divert attention from the objectionable exploitation of resources which makes it doomed in the first place. Being doom could be either due to threats of extinction, market forces or the inherent nature of the society. In Nigeria, it is a large dose of the latter in addition to all the other factors making an effective utilization of crude oil a mirage.

A major shortfall is the lack of corporation between stakeholders. Until the communion of political leadership and civil service execution has more robustness in enforcing equality and fairness, improvements in federal wealth and the fiscal balance of payments is always going to meet severe challenges for filtering down with any degree of uniformity. Decline of the oil industry is global and there are reasons for it. The technical and other features of Nigeria's reserves and what is necessary to bring them to market are unique to the country hence exasperate rather than counter prevailing global malaise of the industry. In 2010 when crude reached a global peak of $140/barrel, the extraction and to market cost in Nigeria was $38/barrel. In 2019, the market cost was around $61/barrel so you can see where the problem is. It is important to point out export revenue in the industry isn't profit.

What happened is that the demise of the industry forced re-development in other areas but it must start from the ground up. That means most of the rejuvenation will happen with other primary/intermediate industries first. Once those rejuvenate, the next step can then be adding value, and locking in more of the value chain on material exported from Nigeria rather than giving away the opportunity to add value so that it can become a generator for other nations at Nigeria's expense. But steps are being taken to put one step before the other rather than taking all steps at a time. This is always Nigeria's undoing. There is lack of political will to build structures and bridges across the oil industry value chain. This then is aligned with other facets of the economy in a manner that promotes endogenous growth within agriculture with the prospect of cushioning revenue when oil windfall winds down.

A strong FMCG sector is developing and exporting to countries in the region in addition to meeting internal demand. This is worth around $30 billion a year and though the oil and gas sector is now around $80 billion, with oil prices as they are, it is hard to see where profit is, all we can now see is turnover. Not only are the other sectors profitizing turnover better, but the flow of money across different verticals and horizontals are convoluted making those industries much harder to embed state sponsored fraud. Except for SNEPCo and SPDC (Shell), there is probably no complete end-to-end extracting major in Nigeria anymore. The majority are just offering different times of engineering and execution services, and money flow is under the auspices of NAPIMS through NLNG and NNPC as vehicles. Error in oil prices history: oil prices continued to be buoyant beyond 2018 with Nigerian crude having a strong market. But in 2020, oil hit a global average for the month of April at $20. This was not entirely unexpected as most mature economies were in deep economic and health crises hence the fall in oil price is in line with low economic activities of the global economy.

Detractors to Building Nigeria's Low Carbon Future

Key actors within the global community have laid out their plan of action to exit fossil fuel before the mid-century. Fossil-led economies must embrace new wave of change if their societies should have sustainable energy future. Clean energy transition viewed largely as a correlate of any workable framework in the efficient attainment of human development. This underscores strong linkages between the Paris Agreement and Sustainable Development Goals of 2015. These global agendas have one thing in common: They are all aimed at reducing poverty, improve energy efficiency, and increase environmental quality ensuring social equity through green growth. Implementation of the NDC and transition to a net zero carbon emission economy is a paradox. Though clean energy is not all about finding new alternatives to fossil fuel per se but is with co-benefits since alternative energy can create employment in rural areas, contribute to foreign exchange savings, and ease dependence on oil as the primary source of revenue. African countries like Morocco and Rwanda leapfrogged in their development trajectory without oil thus becoming a shining light of how to transform an economy or a country without oil but looking towards alternative development pathways. With the immense benefits, uptake of clean energy technology is gaining momentum even though portfolios of constraints still exist. Obstacles to Nigeria's implementation of clean energy transition are multifaceted.

A major reason for Nigeria's complex energy transition is the population of about 200 million with over 120 million living without electricity. Nigeria is sparsely populated hence building electricity grids across long distances to reach communities living on less than $2-$3 a day does not make economic sense. There is widespread energy poverty with most rural communities not linked to the national grid. Likewise, uptake of clean technology and fossil fuel dependent economic system will create significant transformations in the form of employment opportunities for various sectors of the economy.

As noted, Nigeria is reliant on crude oil to grow the economy. Fossil fuel has perilous ramifications for economic growth.[19] Inasmuch as the economy is extensively dependent on a single resource, factors undermining reordering priorities toward undesirable pollution will prevail. With fossil fuel industry's declining profile, deployment of renewable energy will take place at the same time with increases in fossil fuel consumption and production evidenced in the 2016 and 2017 budgets.

Biomass sector is critical to all decarbonization processes in most rural communities. Making a predominantly rural population transit to eco-friendly technologies will have disruptive implication for their livelihoods. In the absence of an appropriate policy framework deployment of biomass-based fuels will eventually exacerbate deforestation causing inadequate replacement of tree cover, soil erosion. Finding an appropriate long term regulatory measure and an energy mix policy with the view to favourably exploring the vast biomass potential will require a new energy optimization policy emphasizing energy efficiency and clean technologies simultaneously ensuring food security.

Cost of renewables is a major constraint to bridging the poverty energy gap. The vast biomass potential is largely untapped because of difficult terrain making implementation of clean energy technology still very expensive. Though prices of renewables are falling globally, it is still above what they can afford. When compared with long term co-benefits of renewables, such cost dwindles in significance. An example of untapped potential due to cost constraint is in hydropower.

Other sources of renewable energy such as wind, hydropower and solar are largely unexplored. Even where renewable energy is minimally produced, the technology is expensive. Finding an appropriate long-term regulatory measure and energy mix to favourably explore the vast clean energy potential will require new policies emphasizing efficiency in water and food production. There is an acute dose of agro-fuel complex with food security imperative. If land grabbed for biofuel production is to flourish unfettered, food/fuel crops contradiction will increase seeing that bioenergy also uses food crops as the feedstock. Government should play a crucial role in resolving energy security impediments redirecting policy and formulating an institutional framework for deployment of clean energy. Political commitment and support of government gives policy direction for the development of necessary regulatory instrument for advancement of the Paris Agreement. Conducive policy instruments are vital components of the clean energy development. Institutional bottleneck is, however, a barrier to sustainable energy management and uptake of clean technologies. Given the policy inconsistency, there are contradictions between the NDC's, 2016 Appropriation Bill and the implementation of the fourth key priorities of the ERGP (ensuring energy security–power and petroleum product). Despite acceding to the global decarbonization goal, the 2016 and 2019 budget committed more resources to energy infrastructure most especially in fossil fuel power generation with insignificant amount earmarked for renewables.

The political class preference is the driving force for failure to institutionalize appropriate building blocks for a low carbon and climate resilient economy to take shape. A good example of a clean vision derailed through neo-patrimonial character of the state is in handling of the clean energy stoves' contract. Under the last democratic dispensation (2015), a contract for N9.2 billion worth of cooking stoves and wonder bags for rural women under the National Clean Cooking Scheme was awarded. This is sound reasoning as biomass consumption would have drastically declined. Such policy's co-benefit is not only in creating jobs when the cooking stoves assembled locally but would have environmental benefits in the form of carbon offsets ultimately earning additional revenue through carbon credits. Although, the project has not fully taken off, it is straddle with all sorts of controversies with graft as the major encumbrance stalling distribution of clean stoves.

With this development, it seems the decarbonization goal of government does not have the political support for limiting emission at all levels. Theoretically, Nigeria's NDC is a bottom-up approach formulated by the politico-bureaucratic elite in consultation with some selected CSOs. One would have expected that national dialogue on an issue like the INDC submission with different ramifications for the nation's economic health of the country is held as it will redirect developmental priorities towards other goals. The want of such consultation means the instrument crafted on decarbonization lacks local anchorage thus forebodes grave danger to Vision 20:2020. Not only will it impede political lock-in on the INDC but will make it is difficult at the micro level to integrate smallholder farmers into the decarbonisation drive.

Dearth of accurate carbon emissions data reflecting the country's carbon footprint is another issue. Data transparency on the actual number of gases flared, ecological imprint and extent of destruction is widespread. Paucity of accurate data is largely an attribute of the lack of capacity to accurately paint in objective terms the environmental pollution generated in all sectors. Given this, most GHG inventory projections in the INDC are based on estimates (or what you may call a wish list) making it difficult to ascertain these claims. Lack of publicly available and transparent data supporting government claims will undermine commitment to exit fossil-led growth. And this is linked to the INDCs inherent deficiency of not laying down in concrete terms a binding commitment as earlier mentioned. The commitments of Paris to end fossil-led development in low income countries contravene the principle of common but differentiated responsibilities and respective capability. Though most countries pledged to limit emissions, conflict abound between national transformation agenda and the global decarbonization framework which is further exacerbate by the energy gap between the North and South.

Climate change is on the front burners of global environmental conversation. Several interrelated issues undermine government's efforts to deal with climate change. One major impediment to the country's climate change mitigation is the vexing issue of paucity of fund. Undue attention is pay inadequacy of fund for meeting developmental needs given that it plays a crucial role of equipping people with the requisite human and non-human resources needed to mitigate climate challenge.

It may seem the political will and commitment at all levels particularly at the highest level of authority to initiate climate change policy is wanting (The national Climate Change Commission Bill is still unresolved). Lack of political will is link to the politically volatile nature with serious ramification for the electoral fortunes of the government in power. Pushing for climate mitigation will have serious consequences for the economic fortune of the country due to the nexus between carbon emissions and economic growth. For one thing, it will have the implication of dwindling foreign exchange earnings if gas flaring is totally prohibited.

Weak legal and policy frameworks on climate change related issues are constraints to climate mitigation. Despite NASPA ratification by the Federal Executive Council (FEC), there is no specific legal framework ensuring climate change issues at various levels. The bill on National Climate Change Commission is enmeshed in a political gridlock resulting from the political horse trade between the executive and legislative arm of government on one hand, and Oil Multinational Corporations (OMNC) and Civil Society Organizations (CSOs) on the other hand. The NASPA remit is to ensure transparent management and allocation of resources in dire climatic situation and this has been either through proactive or reactive responses. In Nigeria, climate change adaptation most especially the instrument of NASPA is approach from the standpoint of necessity in the context of sustainable development along international commitments with greater emphasis placed on acceptable principles that promote economic empowerment, social development and poverty eradication. This constitutes the overriding priorities of government which also explains development blueprints such as Vision 20:2020, APP, NARF, ERGP and other medium to long-term development plans amplifying issues of sustainable development as a prerequisite for effective adaptation governance.

NASPA's impact in terms of advancing a cohesive national adaptation strategy with ameliorating climate change phenomenon as the cardinal goal is minimal. With the formulation of NASPA policy, it is the expectation most impediments to realization of the adaptation goals should have been overcome while climate risk minimized. This is not the case. Instead, there seems to be implementation deficit resulting from issues of integration network between various ministries leading to a co-ordination hiatus. Disconnect between different MDAs left the current adaptation activities as array of policies and programme stand-alone.

The existence of weak institutions to ameliorate climate change related risks is blamed for adaptation deficits recorded in the country. Studies carried out by some scholars bear out how disjointed policies from various institutions at national and state levels are responsible for tackling climate change related devastations but failed to efficiently bring succour to vulnerable communities. Different ministries, agencies of government like the National Emergency Management Authority [NEMA], Nigeria Hydrological Services Agency (NIHSA), the Niger Delta Ministry, OMPADEC, Ministry of Health, Ministry of Environment, Nigerian Metrological Agency, Niger Delta Development Commission, Ministry of Agriculture, Ministry of Water Resources, Ministry of Power, Housing & Urban Development, National Insurance

Commission, Institute for Peace & Conflict Resolution and the Ministry of Niger Delta Affairs are either directly involved in issues of adaptation or indirectly from their secondary impact. Building capacity in health, agriculture, conflict management, energy, housing disaster reduction and water management are primarily post disaster management roles of government. But this must be through coordinated, robust national mechanisms to address the climate induced security challenges.

For now, NEMA in conjunction with the Federal Ministry of Environment is responsible for climate change mitigation and adaptation measures. Another arm of government highly visible on climate change related events is the Department of Climate Change (DCC). The DCC plays a critical role in climate change policy formulation at international arena while the Ministry of Environment carries out not just the supervisory role but are actively involved in implementation activities. Because of their overlapping mandates, some of these organs of government are working at cross-purposes such that there is adaptation deficit and maladaptation in meeting the needs of most societies.

Limited capacity at both institutional and individual to effectively mitigate climate change in these MDAs is a constraint to efficient adaptation implementation. The country is, however, dependent on one-size-fits all policies sketched by the international bureaucracy with little input from the local environment. Lack of local anchorage to harness prerequisite support to propel mitigation action toward achieving the desired objective further stymied adaptation activities. As a result, most efforts to tackle problems of climate change are more reactive than proactive.

Lack of policy continuity is the bane of most policies formulated at national level. High turnover of political office holders together with loss of climate champions have stifled climate action. Turnover rate of proven hands in the specialist field of climate change mitigation may lead to loss of skills and capacity as well as the political will for continuity. Most mitigation activities are implemented by staffs that are inadequately equipped. Hence climate actions are mainly within a few government departments with no serious sectoral coordination. As matter of fact, there is inadequate public awareness on climate change and their potential impact on the social-economic, livelihood and the ecosystem hindering public participation in climate change policy. With inadequate public participation, climate governance is solely the responsibility of government with an insignificant presence from the organized private sector. Owing to threat to shareholders bottom line most mitigation efforts pitched the sector against government. In most cases, if not all, the organized private sector is the ultimate winner due to their strong presence in government.

Like most countries in sub-Saharan Africa, poor data and information on climate change impacts is Nigeria's climate mitigation albatross. Paucity of accurate data on biomass emissions and fossil fuel emission is due to the agrarian nature of the country. Prospects for monitoring greenhouse gases at local and national scales and across sectors is relegated to the consideration of how the country reports statistics. This includes energy and environmental pollution, and policy signals that indicate plans for monitoring systems that are accurate and robust. Even when mitigation action takes place, the infrastructure to actualize and sustain such policies is restricted eroding efficacy of government's efforts to bring about much-desired transformation.

As part of the effort to jumpstart adaptation, government initiated different strategies in consultation with partners such as USAID, DFID, UNDP, GIZ, NEST, Heinrich Boll Foundation, and a host of others. Despite spirited efforts to fashion out some appropriate adaptation strategy impediments to adaptive governance are still prevalent. Coordination deficit leading to conflict and overlapping mandates, dysfunctional arrangements for inter-agency integration, overburden of external (UNFCCC and donor) reporting requirements and inadequate financing for adaptation are still widespread. More so as most of the organs of government lack legal authority, staffing and political support with limited budgets as such failed to address national, regional imbalance and inequalities in global environmental governance. With this, existing institutional machineries are weak thus are not coordinate to ensure they can confront adaptation challenges frontally. The overall effect of institutional failure is their insignificant presence in risk-prone communities since the poor do not have a voice in projects initiated by the institutions delivering services.

More critically, policies formulated to combat climate change are initiated at international level which is not country specific. In some cases, the mechanisms initiated at international level are top-down not bottom-up approach. When policies are externally driven government acts as a rubber-stamping agent of the externally driven agenda. The Paris Agreement attempted to correct such overbearing influence on local communities in embracing a bottom-up approach. Alas, it is also mire in series of controversies consequence of limited local anchorage and political buy-in of major stakeholders. There is also the issue of lack of equity with income distribution limiting the prospects for climate governance because clear majority of the people are still poor; progress in reducing poverty is highly skewed reducing the prospect for translating gains of economic development into social development structures capable of impacting on the ordinary people. Furthermore, gender inequality is high with women and children more prone to climate risks.

Complexities resulting from the ecopolitics of nature's utilization are instrumental to changing relations between nature and humans in the country. Policy implementation is dominated by scholars with free market understanding of nature as primarily one of a large bowl for the take-off of economic growth. Their supposition here is when economic growth has fully taken root, society will develop other alternatives to curtail pollution. Although diverse transformations are afoot driven by neoclassical production model, it has grown out of the hands of its advocates to the level ecopolitics of resource utilization can no longer take the context of nature/man ecological impasse for granted. Evidence of it can be gleaned from the changes in climatic conditions.

Another major challenge is the Post-Kyoto Protocol commitment. Under the new UNFCCC regime, resolving gas flaring controversies involves eliminating carbon emissions. Eliminating carbon belies refocusing the political lens of the country towards an appropriate instrument to replace the drive towards post-Kyoto commitments. Post-Kyoto accord of Paris Agreement is, however, an intricate obligation with answers to improving adaptation potentials of the state while simultaneously curbing carbon intensity of the economy. Balancing the diametrically opposed views requires trade-offs. Trade-offs is inevitable in the conceptualisation of the ultimate paradigm to navigate the murky waters of our eco-politics. It is axiomatic there can be no conservation without first preserving what you have neither can there be preservation in the real sense without conservation. There are critical green areas where the two dimensions are harmoniously interrelated. Their point of departure, however, is the utility value attached hence the need for proper valuation. By utility here we do not mean what is utilised or that ought to be utilised but the effectiveness, efficiency and efficacy of such utilisation.

Against this backdrop, Nigeria must find an enduring pathway in resolving the current ecological unequal exchange. Whether this is really feasible within the prevailing international political economy is entirely a different matter given the market's tendency to redefine its nature. An alternative nature or alternative source of energy amid all these challenges should be the overriding goal of greening the economy. But it raises a new question needing an urgent answer. The question here is not what we are conserving or how we are preserving resources. Rather, it is whether such preservation is an end or a means towards an end?

The Search for an Alternative

The concept of resilience is fast gaining ground as the international norm for preparing people's skill and strengthening their perseverance in a crisis. Resilience is the capacity of social, economic, and environmental systems to cope with a hazardous event or trend or disturbance, responding or reorganizing in ways that maintain their essential function, identity, and structure, while also maintaining the capacity for adaptation, learning, and transformation.[20] Vulnerabilities to climate risks have seriously reduced with resilient subjects' susceptibility to climate change on a decline globally. Vulnerability is the propensity or predisposition to be adversely affected. Vulnerability encompasses a variety of concepts and elements including sensitivity or susceptibility to harm and lack of capacity to cope and adapt.[21]

This is not the case in Nigeria. In Nigeria there is re-appropriation of the concept skewed towards market vision of political life. Although, structural shifts are taking place, it is with a pro-market slant. As of late, transformational process to build a resilient populace is geared towards repositioning market levers as driver of national economy neglecting other dimensions of sustainability. Resilience in the present form has its own demons, however. We experience the reinforcement of resilience demons in the dislocations between sustaining adaptation strategies of the poor and the one-size-fits-all market-led resilience strategies during the 2012 Nigerian floods. Then, reactive measures targeted at carrying out ameliorative actions with limited traction for long-term sustainability. Reinvention of resilience is the fallout of COP21 which made local community implement and plot their ways out of perilous challenges. In line with this adaptation regime, can Nigeria meet her target of unconditionally reducing emission by 20% while also bringing about a 45% reduction conditionally?

The Federal Government of Nigeria also intends to embark on thirteen sectoral activities encapsulated in the National Adaptation Strategic Plan of Action (NASPA) blueprint. Effectively, government intends to embark on deep emission cuts simultaneously reducing vulnerability of the people to climate change. Reducing vulnerability will seriously limit the impact of climate change while also reducing scale of pollution recorded in the country. The NDC presents an opportunity for government to fundamentally shift poverty alleviation and economic development. It is also an opportunity to build capacity of the people to cope with climate risk thus fostering eco-efficient utilization of natural resources. A negative consequence of the Paris Agreement is the reinforcement of existing inequalities in the form of maintaining a strong hold on instruments for building people's preparedness. The mechanism wherein this hold on society is maintained is through the NDC which seeks to integrate social, natural and monetary 'capital' in a continuous framework of power. Here, the engine for mobilizing environmental and financial capital needed for transformational change is the market. Now one may ask: what is resilience?

Generally, resilience is a complex process involving interactions between diverse systems with all exerting influence over adaptability of the system. It is a dynamic process for 'bouncing forward'[20] and changing to a new, more sustainable state. In a polity where different forces hold different notions of building peoples' resistance to change; where most of the views of these forces are shaped through interest group inference, political experience and their historical past is the determinant of resilience. Resilience is primarily the property that captures the capacity of entity 'to anticipate, adapt to and recover from the event such that it resumes its original configuration, shape, functional relationships or trajectory afterwards'.[21] Resilience theory addresses practical and theoretical possibilities for adaptation to climate change as the concept is foundational to contemporary thinking on climate change adaptation. When applied to the problem of climate change adaptation resilience becomes a strategy for administration of life on a planetary scale; making 'resilience' to operate as a *lingua Franca* of risk, preparedness and survivability across the physical, natural and social sciences.[22]

With the concept of resilience in the political domain there is an emerging form of governmentality through resilience. Resilience discourse marks a break with the modernism of 'the 'risk society' introducing a novelty, adaptation, unpredictability, transformation, vulnerability and systems into a governmental discourse that now makes the governance of uncertainty and unpredictability a hallmark of rule'.[23] Climate change is expected to affect hazard frequency and magnitude which represents an increase in disaster risk. The upsurge in risk made it important to mainstream climate change adaptation to Sustainable Development Goals at all levels. So, where there is climate-related disaster, it is indicative of the failure of sustainable development. What this means essentially is the level of development is not sufficient to ensure adequate and sustainable response to climate change. This explains why the most attractive adaptation strategies are those delivering added developmental benefits in the near term, as well as the reduction of vulnerability in the long-term. Various forms of adaptation and resilience are being mobilised to facilitate 'archetypal governmental technologies of neoliberalism; government at a distance, technologies of responsibilisation, and practices of subjectification that produce suitably prudent autonomous and entrepreneurial subjects in a world of naturalised uncertainty and crisis'.[24]

Nigeria receives extensive benefits from the natural environment in the form of goods and services provided for economic growth, energy, protection from floods and soil erosion. Nigeria's expectation is at the earliest stage of development, the country will depend heavily on her abundant natural resources for economic growth. In other words, Nigerian growing economy consists of high levels of throughput from the natural environment and this could be in the form of natural goods and services. In the wake of heightened economic activities culminating in growth and development, natural capital has been increasingly transformed into physical, human and social forms of capital. The transformative process is an attribute of a growing economy in its attempt to rebalance itself. For the economy to develop sustainably, it expects productive structures be flexible enough to allow substitution between scarce and abundant forms of capital. Additionally, there should be a change in technology before productivity increases in the face of declining resources throughput.

Proponents of biomass economy within the country see agriculture's effectiveness in transforming the economy from fossil dependence to a low-carbon economy where carbon consumption causing climate change is minimal. Those who hold this view remit to promote non-fossilized biological materials be used for feedstock in bioenergy production. Effectiveness of biomass as a transformative agent is, however, dependent on those applying biomass technologies and their goal of attaining the bio-based economy. If the bio-economy places a higher premium on shareholders returns, then profit-maximizing technologies will further ensure ecological deteriorations. Then, it is a disservice to the green infrastructure. When the goal of a biomass economy outrightly considers planetary heritage, emphasis will be on efficiency in resource utilization leading to eco-equilibrium between consumption and production. Such utilization should align consumption towards biomass which is less carbon intensive or emit less GHG. Bamboos can guarantee both energy efficient developments while ensuring economic growth. A point worth noting in the transformation of biomass economy is it is sustainable to the level to which we allow green technologies relate to human and non-human nature. If the economic pillar of sustainability superimposed over the ecological or social pillar, the goal of bringing about an equitable transition to a low-carbon society using biomass sector will be defeated. Then, we shall arrive at an eco-disequilibrium wherein economic needs and ecological resources are discordantly interrelated.

In this green regime, land grab is the future of biomass economy. Land grab is an international phenomenon associated with the biomass economy.[25] It is a global phenomenon dealing with purchase of vast swathes of land by wealthier food-insecure nations and private investors from mostly poor developing countries to produce crops for export. For this reason, we can assume on the surface land grabs are beneficial to both foreign investors and host communities alike. Interestingly, some of the current participants in the 'new scramble for Africa' also experienced colonial rule. A major driver of Large Scale Land Acquisitions (LSLA) is the economic hegemonic aspirations of some countries than in claims of feeding teeming hungry population of the South. As such, the search for an alternative source of food and energy is theoretically the main basis for land grabs.

Geo-strategic machination of the investors is the most important motive for land acquisitions and the medium such land deals are guarantee is through use of force. Force, here, ensures compliance. Again, forceful eviction of the real owners of the land is ecologically entropic because further constraint is place on the green infrastructures not just from bio-corporations, but most importantly, the displaced farm families. Resolving this ecological disequilibrium requires curbing depletion from all parties engaged in the decimation of natural capital. We shall presently examine the NDC efficiency as a tool of building Nigeria's climate resiliency.

To start with, an average of 400 million tons of gas is emitted annually from oil exploration and exploitation activities while economic activities from other facets of the oil industry also contribute to GHG emissions. Exploration of oil is noted to have been one of the man-made agents of global warming. In the oil producing region, gas escapes from oil wells in large quantity to the atmosphere. Gases flared have their harmful impact seeing that it contains high amounts of carbon dioxide. A total of 1000 Standard Cubic Feet (SCF) of gas is produce with every barrel of oil while the OPEC oil quota is about 2 million barriers per day. Nigeria produces 2.0 billion SCF of associated gas daily. With an average price of crude oil at $70 a barrel, the projected revenue is in the region of $140 m daily. With gas flaring accounting for 76% of the total GHG emission, it should be a major medium for reduction of the country's emission burden. A 20% reduction of GHG emissions tantamount to about $28 m decrease in revenue generated daily with the attendant consequence felt cross all sectors of the economy.

Government's plan of building resilience of the people is threatened if revenue earnings from this sector decreases. With the uncertainties in crude oil price in the international market, the severity of Nigeria's dire condition becomes more understandable. Government adopted a two-prong approach of pursuing fossil fuel-based growth while also embarking on deep decarbonisation pathway. Taking this path to a carbon neutral future is feasible but promoting carbon dependency in the 2016, 2017, 2018, 2019 budgets will contribute additional adverse carbon load than was committed at Paris. The 2017 budget was to reinforce capital expenditure of the 2016 budget in promoting fossil fuels power generation. A difficult problem is posed here by this position: Taking a strong stance on carbon emission specified in the NDC or promoting economic growth. It will lay the country open to condemnation from other parties to the Paris Agreement. Curbing emission is somehow linked to the political fortune of Buhari's Administration. A severely crucial burden is place on government on the option to take. While it is possible to uphold carbon neutrality to an acceptable level, but some extraneous forces are aligned against attainment of this objective. Although it is feasible to carry-out some readjustments within the next window of opportunity (2020), even then, the options available are becoming limited.

Examining the impact on rural communities within the Niger Delta region will bring out the severity of the problem. Within these communities, rebalancing GHG will be a remarkable development provided it reduces hazards posed polluting activities of oil firms. Gains of improved air quality, health and higher agricultural yields will improve the living conditions of the people. If that be the case, then, it is easy to lapse into political rhetoric without commensurate political will to adhere to promises made. The attempt at implementing the NDC without first enshrining it in the national development plan means it is prone to political interference with the tendency of subjecting a large proportion of the people most especially women and children to more sufferings in the years to come. The promises of Paris will have little or insignificant effect on building resiliency of the ordinary man.

A major reason for the maladaptation within the Niger Delta region is the colossal nature of the economy where there is skewed towards a single resource. Nigeria is guilty of relying solely on one strategic mineral resource not only susceptible to shocks but it is also a source of stress. Curbing emissions from oil industry is a major requirement for effective implementation of the NDC. It is a threat to economic and political wellbeing of present administration. A way out of this dilemma is to fashion an energy mix with serious emphasis on alternative source of energy. This is with the view to safeguarding the wellbeing of present and future generations. The National Climate Change Bill should be given higher priority to achieve this goal. The PIGB is also relevant to address decades of environmental despoliation, wanton pollution and poverty associated with exploitation in the Niger Delta. Owing to the importance in national development, the PIGB will continue to generate extensive controversies. These same controversies now trail the national climate change bill given their crossing cutting impacts on the decarbonisation drive. There is no political will to surmount challenges associated with passing these bills. Uncertainties surrounding passing these bills have serious bearing on the NDC implementation.

No nation can sacrifice the economic wellbeing of her people without strong resistance from some quarters. In most cases when there is the option of choosing between human and non-human nature, the latter ultimately loses out. What we are witnessing today is the trade-off between near-term gains of economic development and long-term protection of ecological assets. A worst-case scenario will be railroading of the NDC into other pressing priorities hence it cannot build resiliency of the people in post-Paris era.

Biomass provides the dominant source of energy with over 80% of the population reliant on biomass for their energy supply. Biomass exists in abundance with vast area with immense prospective. This sector accounts for 24% of the GHG emission in the country. The dominant form of biomass consumed is fuelwood used for domestic purposes. In Nigeria's NDC, the projection is for curtailing charcoal consumption owing to her depletion of forest resources. Large population of Nigerians who depend on biomass would be seriously affected by this commitment. Probably the single thing limiting fuelwood consumption will achieve is the short-term gain of meeting the political commitments of Paris.

In real term terms without adaptation finance biomass consumption will continue to rise. Government cannot proscribe fuelwood consumption without putting in place an alternative. If such measures are implemented without alternative source of energy, government's action will criminalize a clear majority of the people who depend on biomass for their energy supply. To deliver large scale adaptation benefits while avoiding eco-costly deforestation government must plot a pathway out of the country's energy poverty. The plan of action in the NDC to limit using charcoal for cooking is not really feasible within the prevailing economic circumstance of the poor. Putting an end to charcoal consumption will not be determined by what government values most but on how the ordinary people are empowered to key into the decarbonization policy. For now, there is no such luck in the horizon. Transition to net zero economy based on Paris promises will not foist climate resiliency if we follow the current BAU trajectory. Redress can only come if government takes proactive steps in translating political promises into concrete actions in the form of empowering the poor, providing alternatives to charcoal and enacting eco-friendly policies on renewables.

Resilience subject in Nigerian ecological space is in a flux. With the localization of adaptation and associated internalization of coping strategy, the journey to unstable polity has commenced in earnest. The quest to find ecological neutrality in a changing world will be marred by uncertainties of the localization of adaptation. By making a case for localised adaptation the much desired decarbonisation goal of the global economy as a collective strategy face an uncertain future. The goal of building a low-carbon lifestyle based on country funded adaptation actions will reinforce existing silos between human and non-human nature. Resiliency espoused in Paris will not bring lasting solution to some of the biggest challenges of our times but will create new impetus for depletion of natural resources. Such transition requires market tools to correct problems it created, and where market tools are predominant, the net benefit to capital is much higher than those to local communities.

Ultimately, then, a green future demands not just green consumers but green citizens and green policies. The global community's drive toward energy efficient, sustainable and climate friendly economy needs more than just reinventing nature as an economic artefact. Sustainable resource use must aggregate different values, a synergy and trade-off of values balancing the best values from diverse societies to bring about equitable distribution of resources. A new roadmap for power sector reform should be implemented in so doing adopt best practices vital for energy efficiency and conservation in achieving the national decarbonization objective.

Crucial elements of the people's livelihoods must also embrace the efficiency regime for green society to take root. Rebalancing livelihoods of smallholder farmers such that they reduce waste and eco-costly extraction should be the stabilization goal in a carbon neutral world. To achieve this goal, biomass is an integral part of the national development plan with emphasis on priorities toward eco-sufficient alternatives. Emphasizing efficiency will not totally build resiliency. As such, a form of eco-tax should be introduced as disincentive. Disincentives in the form of eco-tax and eco-labelling built into fossil production and consumption makes it less attractive to the fossil fuels industry to increase carbon upload. The envisaged eco-tax should shift the burden from labour to end users and producers.

Capacity to implement the NDC is limited due, in large part, to the lack of financial and technical resources. Awareness of the NDC should be created through incorporation of environmental education into the academic programme at all level. In other to achieve the ultimate goal of reducing dangerous anthropogenic climate change, there must be synergic integration of the NDC with the Vision 20:2020 development policy which further dovetails to NASPA and other national priorities such as the Sustainable Development Goals and other poverty reduction schemes. Because proactive adaptation is critical to post-Paris development, government should develop adaptation strategies on regional level to include cross-border resources management and migration. Prioritization of adaptation strategy should aim at increasing the importance of building resilience of the vulnerable class through proactive strategies that speaks not only to challenges of the rural poor but also addresses adaptation needs of the urban dwellers.

Notes

1. Okoh, (2017) Okoh, A. I. S. (2017). *Pathways to a Green economy in post-Paris Agreement Africa.*
2. Eleri, E.O.; Ugwu O., and Onuvae., (2012) Expanding Access to Pro-Poor Energy Services in Nigeria. A consultation Paper prepared by iced with support from Christian Aid.
3. Heinrich Boll Foundation and Friends of the Earth International. (2015) Coal Atlas: Facts and Figures on a Fossil Fuel Available at: www.Nigeriacoal.org.
4. Ibid.
5. Okoh, A. I. S. (2019) End of Fossil Fuel Economy: Weathering the Storm of Transition to Post -Fossil Society in Nigeria. World Review of Science, Technology and Sustainable Development. Vol. 15,(3) pp.197–213.https://doi.org/10.1504/WRSTSD.2019.102110.
6. Kantai, P. (2011, November 23). Reform of the domestic oil market is sticky business. *Financial Times, Special Report*, p. 2.
7. Ibid.
8. Corruption in Nigeria Dragon-slayers wanted. (2011, December 3). *The Economist*.
9. Shoremeku, K. (2013) Nigeria, Oil and the Yamani Syndrome*Covenant University Public Lecture Series. Vol. 2, No. 2, October, 2013.*
10. Ibid
11. Bill Gates (2017).
12. Karl, T. N. (1997). The Paradox of Plenty: Oil booms and Petro-States: Berkeley: University of California, 1997 p. 342.
13. Ibid.
14. Karl (1997).
15. Sanli, B. Book review: The Paradox of Plenty: Oil Booms and Petro-states Terry Lynn Karl Berkeley: University of California Press.
16. Jeffrey D. Sachs and Andrew M. Warner (1195) National Resource Abundance and Economic Growth. Development Prospects of States Ithaca, NY: Cornel University Press p. 300.
17. The Tribune online Curse of the black gold ·How oil wealth has made Nigeria poor August 14, 2016 Special Report.

18. Ibid.
19. Gavin Wright and Jesse Czelusta (2004). WHY ECONOMIES SLOW: The Myth of the Resource Curse. Challenge, vol. 47, no. 2, March/April 2004, pp. 6–38.
20. IPCC (2014) Summary for policymakers. In: Climate Change 2014: Impacts, Adaptation Part A: Global and Sectoral Aspects. Contribution of Working Group II to the Fifth Assessment Report of the Intergovernmental Panel on Climate Change [Field, C.B., V.R. Barros, D.J. Dokken, K.J. Mach, M.D. Mastrandrea, T.E. Bilir, M. Chatterjee, K.L. Ebi, Y.O. Estrada, R.C. Genova, B. Girma, E.S. Kissel, A.N. Levy, S. MacCracken, P.R. Mastrandrea, and L.L. White (eds.)]. Cambridge University Press, Cambridge, United Kingdom and New York, NY, USA, pp. 1–32.
21. Ibid
22. Welsh, M. (2014). Resilience and responsibility: governing uncertainty in a complex world. *The Geographical Journal, 180*(1), 15–26.
23. Cooper & Walker, (2010).
24. Welsh, M. (2014). Resilience and responsibility: governing uncertainty in a complex world.
25. Cotula et al. (2009) see Land grabs as another avenue to reinvent acquisition of land. To them, traditionally, agricultural value chains tended to concentrate funds in processing and distribution, while the risk fall mainly on primary production, acting as a disincentive for investment in agriculture. But all this is changing with the mad rush for land. The recent traction of land grab is hinged on the upward trend in commodity prices tipping the balance by increasing the downstream risks to processors and distributors involved in sourcing raw materials and boosting returns from production. This increases the attractiveness of investing in agriculture production prompting further acquisition of land and the consolidation of other biofuel-related enterprises.

Chapter 5
Making the Post-vision 20:2020 and NDC Blueprints Efficient as Phase-Out Strategies

"So here's my prediction: You tell me the price of oil, and I'll tell you what kind of Russia you'll have. If the price stays at $60 a barrel, it's going to be more like Venezuela, because its leaders will have plenty of money to indulge their worst instincts, with too few checks and balances. If the price falls to $30, it will be more like Norway. If the price falls to $15 a barrel, it could become more like America—with just enough money to provide a social safety net for its older generation, but with too little money to avoid developing the leaders and institutions to nurture the brainpower of its younger generation."
Will Russia Bet on Its People or It's Oil Wells?
– Thomas L. Friedman, New York Times, February 16, 2007

Preparing for the Post-fossil Fuel Era

Nigeria is experiencing several interrelated challenges with severe consequences for national development. These challenges have their root in the dwindling fortunes of the country which is equally traceable to the decline in oil price and associated decline in crude oil earnings. Sabotage of oil export terminals in the Niger Delta also has negatively affected government's revenue and export earnings, as well as the fiscal capacity to prevent the economy from contracting. With the decline in revenue, the capacity of government to meet developmental needs of the people using foreign exchange accruing from the oil sector has diminished. Spending on much needed infrastructure was equally constrained by the lack of fiscal buffers to absorb the shock as well as leakages of public resources due to corruption and inefficient allocation of capital in the recent past.

In recognition of the magnitude of the fiscal challenges it poses, the Buhari Administration made concerted efforts targeted at tackling these challenges in doing so sought to change the national economic trajectory in a fundamental way. The earliest action taken was prioritization of three policy goals: tackling corruption, improving security and re-building the economy. The Strategic Implementation Plan (SIP) for the 2016 Budget of Change was developed as a short-term intervention for

A. S. Okoh, *Oil Mortality in Post-Fossil Fuel Era Nigeria*,
https://doi.org/10.1007/978-3-030-60785-2_5

this purpose. Interestingly, some visible achievements have been recorded in both economic and ecological tuffs. It is, however, recognized that more needed to be done to propel the economy towards sustainable and accelerated development.

Another major blueprint of government is the Economic Recovery and Growth Plan (ERGP).The ERGP which elapses this year (2020) is a Medium-Term Plan for 2017–2020 which builds on the SIP but is with the objective of restoring economic development by leveraging the ingenuity and resilience of Nigerian people to bring about speedy recovery of the economy. The ERGP is based on the understanding that government in the twenty-first century must evolve from an omnibus provider of citizens' needs into a force for eliminating the bottlenecks impeding innovation and market-based solutions. With the plan coming to an end this year there is strong evidence that the goal of leveraging Science, Technology and Innovation (STI) to build a knowledge-based economy has failed. Critically, the ERGP which is consistent with aspirations of the Sustainable Development Goals (SDGs) and Vision 20: 2020 given that the initiatives address the three dimensions of economic, social and environmental sustainability issues. Still, minimal success has been recorded in these sustainability perspectives.

The ERGP blueprint is a plan of action on recovery of the economy which is different from previous plans in several ways[1]:

First, focused implementation is at the core of the delivery strategy of the Plan over the next four years. More than ever before, there is a strong political determination, commitment and will at the highest level. Whilst all the MDAs will have their various roles in implementing the Plan, a Delivery Unit is being established in the Presidency to drive the implementation of key ERGP priorities. The Ministry of Budget and National Planning will coordinate plan-implementation and for this purpose will, amongst other things, build up its capability for robust monitoring and evaluation.

Second, the Plan outlines bold new initiatives such as ramping up oil production to 2.5mbpd by 2020, privatizing selected public enterprises/assets, and revamping local refineries to reduce petroleum product imports by 60% by 2018. Other initiatives include environmental restoration projects in the Niger Delta, which demonstrate the Federal Government's determination to bring environment sustainability to the forefront of policies. As part of this Plan, oil revenues will be used to develop and diversify the economy, not just sustain consumption as was carried out in the past. The economy will run on multiple engines of growth, not just the single engine of oil. The Plan focuses on growth, not just for its own sake, but for the benefits it will bring to the Nigerian people. This Plan also places importance on emerging sectors such as the entertainment and creative industries.

Third, the ERGP builds on existing sectoral strategies and plans such as the National Industrial Revolution Plan, and the Nigeria Integrated Infrastructure Master Plan. Rather than re-inventing the wheel, the ERGP will strengthen the successful components of these previous strategies and plans while addressing challenges observed in their implementation.

Fourth, the ERGP is innovative in that it signals a changing relationship between the public and private sectors based on close partnership. In implementing the Plan, the Government will collaborate closely with businesses to deepen their investments in the agriculture, power, manufacturing, solid minerals and services sectors, and support the private sector to become the engine of national growth and development. In addition, science and technology will be effectively harness to drive national competitiveness, productivity and economic activities in all sectors.

Fifth, the Buhari Administration merged the Budget and Planning functions into one Ministry to create a better and stronger link between annual budgets and the ERGP. This facilitated the ERGP's preparation process and will also expedite its implementation. It also strengthens the macro framework which underpins the ERGP, ensuring that budgets are properly align with planning, thus promoting effective implementation.

Finally, the ERGP provides for effective collaboration and coordination with the States to ensure that the Federal and State Governments work towards the same goals. The States have a significant role to play in the success of the ERGP and some have already adopted several the initiatives promoted in this Plan.

The Outcome of ERGP

The overall goal of government for articulating the ERGP is one of sustained economic activities and inclusive growth. It hinges on the need to drive a structural economic transformation with an emphasis on improving both public and private sector efficiency. According to government, the ERGP seeks to increase national productivity and achieve sustainable diversification of production, to significantly grow the economy and achieve maximum welfare for the citizens, beginning with food and energy security.[2] This Plan is a pointer to the type of Nigeria the people desire in the short to medium-term, and encourages the use of science, technology and innovation to drive growth. It also provides a blueprint for the type of foundation needed for future generations and focuses on building capabilities of the youth of Nigeria to be able to take the country into the future. But has the plan achieved the desired goal? How has Nigeria fared in key goals of the blueprint.

Agriculture and food security—Agriculture contributed to GDP growth in Nigeria in a consistent manner. The sector grew by 4.88% in Q3 2016 and with as much as 13 per cent in the previous year suggesting immense unrealized potentials.[3] Contray to expectations, investments in agriculture did not guarantee food security with the potential to be a major contributor to job creation capable of bringing about saving on foreign exchange required for food imports. It is projected that successful agricultural harvests will curb inflation and promote economic diversification. On this ground, the ERGP focuses on the needs of the people, prioritizing food security as a critical national objective with plans already in place for national self-sufficiency in rice by 2018 and wheat in the 2019/2020 fiscal year. Unfortunately, the ERGP could not meet this particular target though created awareness for local production of food and reduced the food imported into the country.

Energy and economic growth—The Energy sector is still a sour spot though the ERGP sought to fundamentally ensure development across all other sectors of the economy. For this reason, the ERGP seeks to address issues of energy from the perspective of electric power and the petroleum sector. To ensure adequate power supply, the ERGP aims to optimize delivery of at least 10 GW of operational capacity by 2020 and to improve the energy mix including through greater use of renewable energy. This goal has not been met as power supply is epileptic. Government intended to increase power generation by optimizing operational capacity, encouraging small-scale projects, and building more capacity over the long term. Government intends to invest in transmission infrastructure to curb distributional deficits thus reduce incidence of power outage. There are limited evidences on round of providing these infrastructures. With regard to the oil and gas sector, the intention is to increase the production of crude oil and gas while adding value in the downstream petroleum sector cannot be said to be a success but was derailed by economic recession occasioned by the COVID-19 pandemic.

Urgently increase oil production—The plan to restore production to 2.2 mbpd in the short term and 2.5 mbpd by 2020 to increase export earnings and government revenues by an additional N800 billion annually was not attainable due to the coronavirus pandemic.

Expand power sector infrastructure—Government planned to optimize the delivery of at least 10 GW of operational power capacity by 2020 to boost economic activity across all sectors and improve the quality of life of the citizenry which not budgeted for in all budgets since 2015 but translation of intent into action is still a major problem. Government has not boosted local refining capacity for self-sufficiency and has not reduced petroleum product imports by 60 per cent by 2018, become a net exporter by 2020, save foreign exchange and prevent reversion to the fuel subsidy regime. Rather, the reliance on oil import is growing from strength to strength. The most tangible effort is a private initiative to build the largest single train refining by Dangote group of companies to refine crude oil in Nigeria in 2021.

Transportation Infrastructure—Nigeria's transport infrastructure stock is still inadequate for the size of the economy and constitutes a major cost and constraint for both large and small businesses. Investments in strengthening Nigeria's infrastructure is low hence has not made any significant contribution towards building a competitive economy. Given the scale of the investment required, partnering with the private sector has not taken shape.

Industrialization focusing on Small and Medium Scale Enterprises—Strengthening of small-scale businesses and the promotion of industrialization is priorities for economic recovery. Nigeria's manufacturing sector is still vulnerable to the stagnant economic conditions. It contracted by 4.38 per cent in Q3 2016 largely due to the difficulty of accessing foreign exchange to import intermediate goods and raw materials, and falling consumer demand. This contraction has reoccurred in 2020 as a result of the COVID-19. Infrastructural bottlenecks and an uncompetitive business environment is the main culprit yet, government paid insignificant attention to this pressing issue. The sector was expected to contribute to growth in the short term through policies to improve the usage of existing capacity, through increased availability of foreign exchange and greater domestic value addition. One major strategy was to accelerate the implementation of the National Industrial Revolution Plan (NIRP) through Special Economic Zones (SEZs). The focus will be on priority sectors to generate jobs, promote exports, boost growth and upgrade skills to create 1.5 million jobs by 2020. This plan has not come to fruition with Special Economic Zones receiving no political buy-in from government and private sector. The failure of exporting zone located in Calabar is still fresh in peoples' memories. This grandstanding commitment of creating jobs featured prominently in the NESP document.

Similarly, revitalization of the manufacturing sector will create jobs, stimulate foreign exchange earnings and grow micro, small and medium enterprises (MSMEs) came out in the NESP document only it is more aligned towards SMEs. The involvement of small businesses in the service sector is a major lever for economic recovery but just as this sectors did not recover in the ERGP we are yet to see what has changed for the NESP to usher in the desired recovery.

The ERGP will end in the coming four (4) months without significant actions in the economic arena. The major gainer is ensuring recovery of growth and placing agriculture in the front burner of national discuss. Though there is some form of transformation within the rice producing areas which brought about interest in local rice. But all these gains came to an inglorious end with the COVID-19 pandemic which truncated the narrow gains in rice production. To build on this, the President in April, 2020 constituted a new committee to sketch a post COVID-19 recovery plan for the nation.

NDC and Nigeria's Decarbonisation

At COP21 a total of 195 countries met in December 2015 pursuance of the UNFCCC objective of limiting rise of the mean global temperature to 1.5 °C and below 2 °C of preindustrial level threshold. The instrument used for submitting country specified GHG reduction targets at COP21 was through a new instrument known as the Intended Nationally Determined Contributions (INDCs). The INDC is a country specific decarbonisation target representing a progression over time of GHG reduction to be implemented within a state by all the parties to Paris Agreement. One single tangible objective of the conference is ensuring collective aggregation of individually submitted INDCs will not only radically limit global warming but will rebalance consumption and production of carbon globally while ensuring sustainable development, poverty reduction and food security. The INDC also serves as a plan of action on how diverse countries will adapt to the adverse impacts of climate change simultaneously fostering a climate resilient and low carbon development in a manner it does not threaten their food production.

Nigeria's INDC submission at Paris Conference followed the general trend in Africa of maintaining a common position on issues pertaining to climate negotiations. However, through Nigeria's submission followed similar pattern with other African countries but is slightly different slanted toward the country's biogeophysical and economic realities. The INDC has mitigation and adaptation components:

I. **Mitigation**

Nigeria's INDC is to make an unconditional contribution in terms of reducing GHG emissions of 20% below BAU consistent with the current development trends and government policy priorities. Policies and measures to deliver these savings include: improving energy efficiency by 20% unconditionally, with 13 GW of renewable electricity provided to rural communities currently off-grid and ending gas flaring. Without ambitious mitigation action, Nigeria's per capita emission is to grow to 3.4 tonnes CO2e. Given the slow pace of development, emission from conditional contribution (45%) is to stabilize slightly above today's level of 2 tonnes per capita once the full package of policies and measures has been implemented with international support.

According to government, the *Nigeria Climate Change Policy Response and Strategy* (NCCRS) was adopted in 2012 and the National Adaptation Strategy and Plan of Action for Climate change in Nigeria (NASPA-CCN) in 2011 to ensure an effective national response to the significant and multi-facetted impacts of climate change. The strategic goal of the *Nigeria Climate Change Policy Response and Strategy* is to foster low-carbon, high growth economic development and build a climate resilient society through the attainment of the following objectives[4]:

- Implement mitigation measures that will promote low carbon as well as sustainable and high economic growth;
- Enhance national capacity to adapt to climate change;
- Raise climate change related science, technology and R&D to a new level that will enable the country to better participate in international scientific and technological cooperation on climate change;
- Significantly increase public awareness and involve private sector participation in addressing the challenges of climate change;
- Strengthen national institutions and mechanisms (policy, legislative and economic) to establish a suitable and functional framework for climate change governance.

Other measures identified in the INDC are:

1. **Energy**

- Renewable energy, particularly decentralized
- Multi-cycle power stations
- Scalable power stations of 20-50 MW
- Enforced energy efficiency
- Use of natural gas rather than liquid fuels

2. Oil and Gas

- Improved enforcement of gas flaring restrictions
- Development of Gas-to-Power Plants at Gas Flare Sites (micro grid)
- Blending 10% by volume of Fuel-Ethanol with Gasoline (E10) and 20% by volume of Biodiesel with Petroleum Diesel (B20) for Transportation Fuels.

3. Agriculture and Land Use

- Climate Smart Agriculture
- Stop using charcoal

4. Industry

- Benchmarking against international best practice for industrial energy usage
- Adoption of green technology in industry

5. Transport

- Modal shift from air to high speed rail
- Moving freight to rail
- Upgrading roads
- Urban transit
- Toll roads/road pricing
- Increasing use of CNG
- Reform petrol/diesel subsidies

The summary of Nigeria's overall mitigation is captured in Table 5.1.

II. Adaptation Component

Adaptation to climate change is as contain in the National Adaptation Strategy and Plan of Action for Climate Change Nigeria (NASPA-CCN) blueprint. NASPA-CCN describes the country's adaptation priorities bringing together existing initiatives and priorities for future action. The central plank of NASPA policy is not just to promote climate change mitigation and adaptation at local level but is a bridge between current disaster risk management efforts and reducing vulnerabilities to climate change. Generally, NASPA policy seeks to minimize risks, improve local and national adaptive capacity and resilience, leverage new opportunities, and facilitate collaboration with the global community, all with the view to reducing Nigeria's vulnerability to the negative impacts of climate change. The goal of NASPA-CCN is to tackle the menace of climate change headlong in so doing proactively implement coping and adaptation strategies for the people. Adaptation measures to curb the impact of climate change in post Paris era inter alia, include:

– Measures to build resilience of the people in the agro-ecosystems to ensure their capacity to provide ecosystem services,
– Ensure the integration of the agriculture sectors as well as food security and nutrition concerns in climate change strategies and policies,

Table 5.1 Summary of key aspects of Nigeria's INDC

Aspect	Detail
Type of objective	Reduction from Business as Usual (BAU)
Target Year	2030
Implementation year	2015–2030
Base data period	2010–2014
Summary of objective	Economic and social development: grow economy 5% per year, improve standard of living, electricity access for all
Unconditional and conditional mitigation objectives	20% unconditional, 45% conditional
Key measures	– Work toward ending gas flaring by 2030 – Work toward Off-grid solar PV of 13GW (13,000 MW) – Efficient gas generators – % per year energy efficiency (30% by 2030) – Transport shift car to bus – Improve electricity grid – Climate smart agriculture and reforestation
Emissions per US$ (real) GDP	0.873 kg CO_2e (2015) [0.491 kg CO_2e (2030)]
GDP per capita (US$)	2,950 (2014) 3,964 (2030; real 2015 US$)
Estimated emissions per capita	Current: around 2 tonnes CO_2e 2030 BAU: around 3.4 tonnes CO_2e 2030 Conditional: around 2 tonnes CO_2e
Global warming Potentials used	IPCC Fourth Assessment Report
Cost estimate data	National Cost = $142b; National Benefits = $304b (World Bank report "Low Carbon Development Opportunities for Nigeria" (2013))
Gases covered	CO_2, N_2O, CH_4
Emissions as % of global total	<1% (2010)
Historical emissions (1850–2010)	2,564.02 million tones

Source FGN (2015).[5]

– Integrated strategy related to water management, land and natural resource management, rural development and social protection.

Criticisms of the Paris Commitment

Reminiscent of the Kyoto Protocol before it the Paris Agreement has its own internal contradictions. From the inception series of criticisms trailed the conference outcome.

One critical issue is in the formulation of the equity dimension of the Agreement. Undeniably, the inter-temporal dimension of climate change mitigation and adaptation are not concretely captured in the present accord. Issue of common but differentiated responsibility and respective capacity is no longer the gauge for a country's exemption from binding emission cuts. The UNFCCC regime focuses on minimizing activities of all polluting industries and countries. In this setting, insignificant attention is pay vulnerability of those handicapped from the effects of the poisonous gases already emitted. The primary focus of this regime is meeting short-term targets against paying due attention to long-term adaptation challenges. Despite remarkable promises of bringing global temperature rise to 1.5 °C/2 °C,[6] controversies trail the Paris Agreement. The non-legally binding nature of the Agreement and no decisive or clear stand on climate finance contribution is another. Though the Agreement may not be perfect as would have been expected, however, it is a step forward in efforts to tackle climate change. The phrase "in the light of national circumstances" in the Agreement is subject to various interpretation by parties signatory to the Agreement. This has created an escape route for heavy polluters and diminished the basic Principle of Polluters Pay concept thereby putting additional burden of servicing the climate debt on most of the developing countries still grappling with other developmental challenges and which contributed the least to the problem.

One of the most virulent criticisms it faced is from the father of climate change awareness: James Hansen. James Hansen, a former Director of National Aeronautics and Space Administration (NASA) in commenting on the outcome of COP21 Hansen criticized the agreement for reinforcing carbon emission rather than reducing it. His criticism is hinge on COP21 attribute of promoting carbon emissions. To him, the Paris Agreement cannot deliver on the promises because if fossil fuel appears to be the cheapest fuel out there they will continue to be burned. In Nigeria and other OPEC member states, fossil fuel is not just the cheapest source of energy it is also the main source of revenue. According to Hansen, the Agreement lacks any concrete action on how to curtail emission thus cannot bring about any significant change in the mean global temperature. (This very criticism is significant due, in large part, to extraction of fossil fuel which is making it difficult for some countries to concretely state their emission reduction and the actual quantity from the oil industry).For this reason, the Paris Agreement is merely 'worthless words' in capable of easing emission instead will push the world toward further carbonization.[7] On this ground, the conference is a 'fraud' on a worldwide scale perpetuated by world leaders with active connivance of the captains of industry.[8]

Again, this agreement amounts to 'words without sword' decried as mere 'sound bites' which cannot get its foot on the door of climate change. To these critics, the Paris declaration of intent is an open-ended commitment. For this reason, climate action wrought at Paris will go the same way as the Kyoto Protocol. The result is, some countries will pull out of the Agreement even before the first commitment period of 2020 under similar circumstance when USA, Russia and Japan repudiated the Kyoto Protocol before it. A case in point is USA intransigence in ratifying the terms of the Paris Agreement creating apathy in countries where the political actors are not fully committed to the goals of the UNFCCC.

The Paris conference on climate change is also considered an 'epic fail on a planetary scale' while others disparage it for institutionalizing corporatists' profit with forest resources turned into another commodity of trade. Since corporatists goals are major determinants of Paris action emission cuts promised are insufficient to bring about the desirable level of 1.5/2 °C global average temperature not only does it promote inequality but will enhance poverty. Nigeria's estimated annual emissions are approximately 90 million metric tons of carbon dioxide equivalents (Mt CO_2e) per year, of which the dominant source is through gas flaring. Other major sources are from on-site use of gas (mainly for power generation) for operating oil and gas production, transportation, and processing facilities; fugitive emissions of gas through leaks and other losses; and venting of gas from oil storage tanks. Furthermore, over 70% of the population rely extensively on biomass energy as their primary source of energy. Most of the poverty in Nigeria is related agriculture. Over 80% of farmers are poor with 87% in core poverty class. A recent research shows an average of six Nigerians join extreme poverty class in every minute of the day causing the country to surpass India as the world poverty capital. Of this, most of the poor live in rural areas and are dependent on biomass energy which contributed to the country's high GHG emission upload.

Challenges of NDC Implementation

The NDC will transform Nigeria encouraging new investment opportunities in clean technology while also reducing atmospheric GHG emissions towards the attainment of the Paris Agreement.[9] Despite this potential; the Paris Agreements has several socio-economic and political encumbrances impeding the implementation. One of these challenges is the issue of marginalization. Marginalization of majority of the people who are poor is a major constraint to achieving objectives of the UNFCCC. What this reveals is that Agreement will have different ramifications for diverse societies. Though the outcome of COP21 would engender growth in some quarters, it will also create or entrench inequality in others. Implication of the Agreement is far more critical in Nigeria where the NDC's commitment under the Paris Agreement could place new limits to growth which then will invariably reinforce inequality.

Obligations of Paris negate the principle of common but differentiated responsibility and respective capability. If followed in objective terms, the NDC will seriously impair the country's developmental goal encapsulated in the Vision 20:2020 blueprint. Furthermore, at COP21 there is declining emphasis on adaptation evidenced in the non-binding component of adaptation finance thus could detract from the global commitment. Though Marrakesh (COP22) tried to bridge this gap but implementation fissure will be a serious issue in the years to come. COP24 attempted to correct the shortfall in the implementation by formulating a rulebook. Still, complexities in the rulebook can derail nationally determined reporting scheme. To resolve this, COP25 focused more on market mechanism which also comes with its own complexity of effective implementation. Without binding commitments, the rulebook will be subject to country political and economic circumstance.

Although the concept of resilience is gaining ground at the international level and is now the recognised norm for building people's skill and strengthening their perseverance in a crisis, still, resilience has its own demons due to the complex nature involving multiscale interactions between diverse systems and stakeholders. These stakeholders hold diverse notions of building peoples' coping strategy in perilous times. Though resilience to climate risk is improving globally, we are witnessing re-appropriation of the concept skewed toward market-led vision of political life. Dislocations between sustaining adaptation strategies of the poor based on the one-size-fits-all market-led resilience strategy which insists local communities implement and plot their way out of climate extreme events emerged.

Unfortunately, uncertainties surrounding fashioning an appropriate blueprint to manage the changing environment of decarbonisation at the same time with sustaining economic development have not helped matters. Fashioning an appropriate economic roadmap to drive implementation of the Agreement placed serious burden on policy implementers. Finding an appropriate economic toolkit to propel society toward the global decarbonisation objective did not receive attention it deserves. Sundry economic toolkits are floating around in political circles to drive the decarbonization initiative. At a glance, it might seem that the world leaders are comfortable with dissimilar economic blueprints existing side by side at national and international levels. Monitoring and reporting implementation in this situation will be weak. The Transparency (of action) Framework in the Paris Agreement still require further elaboration if countries would be held to account. This is another challenge for NDC implementation. Inadequacies of an appropriate implementation instrument and weak reporting regime pose a serious problem to the Paris Agreement. Monitoring, Reporting and Verification (MRV) have been adduced solution to reducing the level of GHG emission.

The onus of reporting compliance with the global initiative rest with the government of Nigeria but ensuring compliance to the Agreement will task an overburdened civil service. Besides that, different states hold varied norms and values hence compliance with the Agreement rulebook espoused at Katowice COP24 could be under reported. In Nigeria, where there is no agency or institution vested with the concrete responsibility of coordinating NDC implementation it therefore means the overburdened Department of Climate Change (DCC) will be responsible for the NDC. Still, the DCC lacks authority and legal backing to institutionalize far reaching prioritization of emissions.

NDC and Post-COVID-19 Recovery

The Nationally Determined Contribution is a country specified decarbonization goal in line with economic circumstance of the state. It is expected to reduce carbon consumption and production while simultaneously enshrining a culture of proactive relationship with nature. It seeks to reinforce government's commitment by regulating the uptake of carbon within the country. Even then, the global decarbonization objective goes beyond the narrow prism of political commitment to a more broader terrain linked to outright GHG reduction drive. The real problem government faces here is not so much that of the implementation of the NDC but more of adjusting successfully to meet new challenges in the socioeconomic space. These new complexities are not only affecting the quality of life but are also eroding the country's capacity to meet her other obligations.

Under the terms of the Paris Agreement, new or updated NDCs were due from each party to the agreement in five-year cycles. These were to be submitted well in advance of Glasgow, but the COVID-19 crisis has thrown a wrench into the process for many countries. Not only are many government staff who were involved in the NDC preparation now focusing on the COVID crisis, but many of the underlying assumptions on which NDCs are based – such as the availability of domestic budget resources, borrowing headroom or access to international climate finance, economic growth and emission trajectories – are coming into question. So say Stephen Hammer & co.[18]

The NDCs is a major avenue to mainstream national plan with international obligation. The NDC comprise both the mitigation and adaptation component and provide a window into the government vision for areas of future economic growth and technology transformation, both of which have clear linkages with job creation. They frequently must account for complex challenges where progress can deliver multiple wins in terms of reduced poverty, enhanced competitiveness, improved public health and local quality of life, and global public goods.

In 2020 with the prevalence of COVID-19 additional work will certainly be required to turn a country's NDC into a pipeline of shovel-ready activities, given that most lack detail on the exact locale and the co-benefits achieved (or risks associated) with each proposed intervention. Current NDCs are not enough to achieve full decarbonization of the world economy and avert the worst of the climate crisis. The coronavirus of 2020 will add additional constraints to the global economy requiring synchronizing NDC with COVID-19 recovery efforts.

Nigerian government announced different stimulus packages with the Ministry Finance expected to coordinate actionable activities in transport, health, energy, agriculture and environment to build a climate resilient society attune to uncertainties in the global economy. This requires concerted efforts from all stakeholders to meet decarbonization goal while ensuring human health. Different stakeholders can advance this effort in various ways, for instance:

- **For government**: Set up a coordination mechanism jointly between ministries that oversee NDCs and the sustainable recovery, to manage the process of project consideration and keep an updated list of projects.
- **For the climate finance community**: Explore and establish ways to make grant support available to help with project preparation, with a focus on converting NDC commitments into actionable projects that would be competitive in stimulus discussions.

- **For the formal climate process**: Encourage upcoming NDC submissions to include country updated list as a measure to enhance cooperation and boost access to resources.
- **For financial and lending institutions**: Take the updated list as a client government's priorities and support them through financial and technical assistance.

Governments around the world have spent considerable time and effort in recent months and years preparing new NDCs that chart a safer and more sustainable future for their citizens. Taking these on board now as part of recovery planning can help us all build back better from the current crisis. Activities to propel growth in post-COVID era should also embrace the NDC and other national priorities such as the new national vision. It covers two timescales: short-term needs to deliver as many jobs, income, and economic demand as quickly as possible, and the longer-term need to deliver sustainable growth and prosperity. Over the short term, there are three main considerations suggested by Hammer& Co[19]:

- *Job creation*, looking at the number of jobs created per dollar invested, but also the types of jobs created and who benefits from them, and the match between the skills needed and those available in the local workforce.
- *Boost to economic activity*, focusing on the economic multiplier each intervention can deliver, the ability of a project to directly replace missing demand, and its impact on import levels or the national trade balance.
- *Timeliness and risk*, assessing whether the project generates stimulus and employment benefits over the very short term and whether they are durable even in the face of possible re-imposition of local quarantine measures.

Over the longer term, a project must also support countries on three different dimensions:

- *Long-term growth potential*, looking at its impact on human, natural, and physical capital. For instance, some projects do better at improving human capital, by building the future skills and health of the population, especially if air and water pollution can be reduced, or access to improved drinking water is improved. Others may promote the use of more efficient technologies, provide important public goods like modern energy or sanitation, or address market failures, such as distortive subsidies that are obstacles to long-term growth.
- *Resilience to future shocks*, with interventions to build capacity for societies and economies to cope with and recover from external shocks, like COVID-19 today, but also other forms of natural disasters and future climate change impacts.
- *Decarbonization and sustainable growth trajectory*, with actions to support and disseminate green technologies, like grid investments that facilitate the use of renewable energy and electric vehicles, or low-tech options like afforestation and landscape and watershed restoration and management. It will be particularly important to ensure that investments from stimulus packages do not impose large stranded asset costs on the economy in coming decades, for instance because they bet on declining technologies or place projects in high-risk flood zones.

Governments seeking to apply this framework may wish to use it in two phases. First, it can be used as a quick cut, "yes–no-maybe" assessment identifying the "worst offenders." The goal is to ensure governments don't invest in projects that are attractive for their stimulus characteristics but detrimental over the long term. In a second phase, the proposed indicators can help decision makers prioritize among any remaining projects, identifying "best in class" projects that deliver multiple benefits to society. Policymakers have a lot on their plate right now, and economic recovery plans cannot move faster than efforts addressing the current health crisis. But as governments shift their focus to recovery, the choices countries make will define what tomorrow looks like and whether we are better able to manage future global crises.

Notes

1. FGN- Federal Government of Nigeria (2017). Economic Recovery and Growth Plan 2017–2020. Ministry of Budget and Planning.
2. Ibid.
3. Ibid.
4. FGN- Federal Government of Nigeria. (2015). *Approved Nigeria's INDC Submision.* Retrieved from https://www.unfccc.int/submisions/INDC/submissionpages/submissions.aspx.
5. Ibid.
6. IPCC 2014.
7. Okoh, (2014) asserts that UNFCCC pledges are not adequate to stem climate risks as global emissions will still exceed plausible trajectory of 2 °C above pre-industrial levels in the years to come. Post-Kyoto Protocol negotiation at COP20 Lima in Peru was deemed a game changer critical for finding deep decarbonization pathway for Paris COP21. Finding a fair and equitable pathway to the future opened deep rooted climate politics. The age-worn Pandora Box of climate politics between developed and developing economies in resolving the greenhouse gases emission cuts gridlock re-emerged at COP20. With carbon emissions reaching an all-time high of 400 ppm in 2014, limiting atmospheric emission to less than 2 °C of pre-industrial levels considered the safe threshold for planetary bounds is vital to negotiations of COP21 slated for Paris.
8. Lingo's conclusion is that Paris Accord is a mirage of seemingly reducing emissions while in reality increasing them. In their words: The overall picture is very disappointing: fossil extraction will continue to increase in most places, in spite of the Paris Accord. It is essential to include the extraction sector in climate change mitigation policy. The amount of fossil extraction over the next 15 years is going to determine whether we will enter the danger zone of likely run-away climate change. Unless we address extraction head on, COP21 will offer a mirage: a picture that looks like reducing emissions while we are in fact increasing them (Lingo, 2015, p. 1).
9. FGN, (2015).Approved Nigeria's INDC Submision.
10. National Bureau of Statistics (2014).

11. FGN, (2015)..
12. Ibid.
13. Ibid.
14. Okoh. A. I. S, Mailumo, D. & Orokpo O. F. (2017). Nigeria's Nationally Determined Contribution (NDC) in the Transition to a Low Carbon Economy. *Dutse journal of Economics and Development Studies (DUJEDS)*. Vol. 3 No. 1. pp. 27–41.
15. Energy Commission of Nigeria, (2008).
16. Bala 2013.
17. Ibid.
18. Stephen Hammer, Stephane Hallegatte and Ferzina Banaji in a World Bank blog of 5th May, 2020 (https://blogsworldbank.org/team/Stephen-hammer).
19. Ibid.

Chapter 6
Roadmap to Nigeria's Future Without Oil

Check for updates

"The Best Way to Predict the Future is to Design It"
– Abraham Lincoln

Diversification as Fossil Fuel Exit Strategy

The shift from fossil fuels will make Nigeria face a bleak future in a short term but with long term benefits. Economic diversification is the solution since it has high political visibility but slanted toward vertical diversification. The policy space face crisis of overcoming the slow pace of horizontal diversification. One would have expected that the pace of diversification of human capital to outstrip the resource sector. Unfortunately, oil and gas sector are the main medium for meeting developmental needs of the people. Till date, pursuit of low carbon strategies and economic development measures exist side by side in the economy. This is due, in part, to the abundant fossil energy reserve. Nigeria is a major oil producing country with proven crude oil reserve of 37.2 billion. Furthermore, the country is blessed with enormous natural gas deposits. Natural gas from the oil producing areas provides power for households and other industrial uses. It is estimated about 50,000 MW capacity driven by natural gas be made available in 2020.[1] On the contrary, only 8017 MW has been put in place. The country also has vast coal deposit of 2.7 billion tonnes although its utilization is negligible because most of the coal producing mines have become moribund. If we must add coal in the energy mix, then it must be clean coal in the light of the various multilateral environmental agreements the country is signatory to and in the face of changing climate due to increase in GHG emissions which is associated with coal exploration.

Nigeria's vast renewable energy is to contribute only 20% to electricity generation through uptake of energy efficient technologies leading to its doubling in 2030. What this means essentially is that alternatives derived from renewable energy are projected to drive the economy. Although the determination to cut the emission intensity of the

A. S. Okoh, *Oil Mortality in Post-Fossil Fuel Era Nigeria*,
https://doi.org/10.1007/978-3-030-60785-2_6

economy is high, there is no clear-cut policy that speaks to what the country's power generation and energy mix can and should be; where such vision is aligned with Sustainable Development Goals, industrialization and climate protection objectives of government. Instead, there are hotchpotch policies with exit from petroleum at heart though lacking a systematic plan of action in achieving this goal.

The biofuel production meets diversification goal the country. The sector receives moderate attention for this reason leading to various biofuel projects located in Benue and Ondo states by the Nigerian National Petroleum Corporation (NNPC). These projects are earmarked to earn additional income through generation of \$150 m annually.[2] A three-pronged approach was developed to ensure the smooth take-off of biofuel production. These three programmes intended to bring about a low carbon transition are: Domestic biofuels industry development, seedling development, biofuels policy development and incentives.

Though there are large coal deposits, all the coal mines have stopped functioning.[3] In realization of the energy generation goal, government has commenced a comprehensive programme of resuscitating the coal industry. Notwithstanding, the oil sector is the main source of industrial energy. Specially, there are 11 million hectares of woodland forest with enormous biomass potential in the country. Traditional biomass is the primary source of energy, hence the energy sector vital for green growth is evolving. The biomass sector provides 65.50% of the primary energy supply with over 70% of the population engage in biomass-related economic activities.

To realize the renewable energy goal, a low scale renewable energy production has taken off with the objective of lessening overwhelming reliance on gas to generate power as well as reduce gas flaring, which made her the second top gas flaring nation globally. To ease relying on fossil fuels, the Federal Government of Nigeria approved the National Energy Policy (NEP). In 2003, the country also articulated the sustainable exploitation and utilization of all viable energy resources such as hydropower, solar energy, biomass, wind and integrated hydrogen. Exploitation of these alternative resources is to be in an environmentally sustainable manner with private sector and indigenous participation in the development activity. Buhari's Administration issued solar energy permits to 14 private entities for generation of energy to meet government power generation objectives contained in the ERGP.

Despite government's commitments, renewable energy development is a high-risk investment owing to the strong presence of fossil kingpins in the policy formulation arena. Consequently, the deployment of renewable energy which should commence with the signing of PPP is enmeshed in serious controversies. Complexities associated with the renewable deployment is unresolved due to the lack of a clear-cut vision where different blueprints such as the 30:30:30, Sustainable Energy for all, Action Agenda National Renewable Energy Action Plan, Rural Electrification and Implementation Plan and NEP compete for attention.

In 2005, over half of the country's emissions came from agriculture, forestry, and land use change. It was expected in the Vision 20: 2020 that the GDP share of agriculture will decline when the remaining forests are depleted. The contribution of this sector to the total emission has not decreased. There is no accurate GHG emission inventory clearly stating what the country's GHG emission debt is. Absence of a clearly defined long range forecasting and generation of privatised electricity with renewable energy as a critical component of the low carbon growth is rampant. It is argued, for instance, that the low carbon growth is a political obligation seeing that the country lacks a cogent clean energy strategy and growth plan like the FES articulated by other oil producing nations such as Brazil where bioenergy will be the main sources of energy used in cars.

Nigeria's hydropower potential is 15,000 MW although only about 2000 MW is currently exploited with private electricity supply companies meeting minimal energy supply. It is the projection that about 14,000 MW of the hydro should be available by 2020 in the country. Solar energy has the greatest potential to contribute enormous amount of low carbon energy with its capacity in the region of 4–6.5 kWh/m2/day. In spite this, unlocking the potentials is still work in progress. In the NEP blueprint, it is estimated about 13,000 MW of off-grid PV solar electricity is put in place in 2030. However, only about 15 MW of dispersed solar PV systems are on the ground. If the country can cover only a small fraction of the landmass with solar panels, it could produce 192 000 MW of power, compared to the 4000 MW currently available on the national grid from diverse sources of energy generation.

More importantly, tapping the vast energy potentials is a serious issue with the Federal Government declaring a gas-to-power revolution. Wind resource of 2–4 m/s at 10 m height is considered low for wind farm electricity generation using conventional wind generations. At a higher height, greater wind speeds may be encountered such that reasonable electricity could be generated. It is envisioned about 40 MW of wind electricity would be needed in 2020. Though oil production is one of the major sources of growth, agriculture still contributes 42% of GDP but crude is a major revenue earner hence putting end to GHG emissions from fossil fuel energy will severely affect the country. That much we have already noted. We have also stated how Nigerian economy is highly susceptible to external shocks; taking a serious bashing when crude oil price crashed in 2015; in so doing, undermining the narrow gains recorded in the last ten years. Remarkable growth achieved in 2014 GDP rebase is a pointer the country was on track to achieve between 7 and 9% growth. Gains achieved have, however, been eroded by insecurity, rising unemployment, failing infrastructure, dwindling revenue from the oil and gas sector and climate extreme events in the form of floods. The situation is far more critical today because two-third of the population who depend on agriculture are vulnerable to climate variability. Rising incidence of climate change most especially the perennial flooding of the Benue and Niger troughs is a threat to APP, ERGP and Vision 20:2020 blueprints. Power generation poses serious impediment to the government policies. Other challenges inter alia are: energy poverty, price of renewables, food and biofuel complex, land use change, investors' apathy, political manoeuvrings from an unchanging oil industry, technological handicap and intermittency shortfall in renewable energy have cumulatively made exit from fossil strategy a complicated venture.

A viable solution exploited in other developing economies is to take advantage of the $60 billion per year estimated total value of bamboo and rattan market. Bamboos have over 10,000 uses and can generate extensive income for the country in the post-carbon era. Sketching a new roadmap to reprioritise consumption and production of hydrocarbon entails fashioning a new blueprint that integrates bamboos not just to the energy mix but strives to gain additional revenue from the export market. Within this roadmap there should be a clear-cut policy that speaks to what Nigeria's electricity generation mix can and should be, where such vision is aligned with the country's sustainable development, industrialization and climate protection goals. For now, what we have are hotchpotch policies with exit from fossil dependency at heart. In the power sector there is, for instance, no clearly defined long term forecasting of generation for privatised electricity with renewables as a critical component of the low carbon growth.

Note: We have already called attention to how there is lack of a coherent, clean energy strategy and growth plan like those articulated in Saudi Arabia. If such plans are properly conceptualized, linkages of bamboos to growth and its ecological benefits will be established. At present, bamboo's implication for sustainable development exist within the framework of INBAR documents government appended their signature to. Until the political membership of INBAR translates into concrete actions, Nigeria will be within the goal of meeting its Vision 20:2020, ERGP together with the NDC commitments but might not realise these development objectives. There is the imperative to de-emphasis fossil fuels which plunged the country into the worst economic downturn towards a post-Oil Age whereby biomass most especially bamboo is made cornerstone of economic development.

Transcending Policy Implementation Gap on Economic Diversification

Diversification of the economy will combat the myriad of problems beleaguering Nigeria today. Finding a new approach to end the reliance on petroleum resources would open the country to tap its vast wealth in other natural resources with the co-benefits of improving the standard of living of the people while maintaining a stable environment. Nigeria is highly susceptible to the vicissitudes of external shocks given that the high-income base constrained its ability to withstand concerted stress. This is an anomaly. The reverse is the case in most economies. Instead of the high income engendering an atmosphere where pressures are overcome, the country is being pushed towards negative change making it difficult to surmount the uncertainties of the oil market. It is ironic a country noted to be the seventh largest oil producing nation is equally one of the poorest in the world; occupying an unenviable position in the Human Development Index (HDI). The World Bank ranks Nigeria 152 out of 157 countries in the HDI. According to the former President of the World Bank Group, Dr. Jim Yong Kim, Nigeria, like many other African nations is in the red zone because the country's health and education budgets are low.

Of note, the petroleum sector is the poster child of all that is wrong with Nigerian economy. Not only because of the large presence of the lethal cocktail of environmentally unfriendly emissions but more of the failure to transform the society as expected. Although over $400 billion have accrued the country since independence as crude oil revenue, it failed to bring about the promised transformation. As Nigerians continue to grapple with the consequences of this failed promise, other options are considered viable solutions based on the historical knowledge of its potency of bringing about the desired economic benefit. It is, therefore, not surprising then that agriculture with its historic significance played an active role in promoting inclusive growth. From the first development plan to date, activities targeted at promoting agriculture as signpost of accelerated development.

Nonetheless, the country is not rooted in long-term concern about economic diversification as such. The country is treading this slippery slope without any tangible outcome; neither was the issue of promoting economic wellbeing of the poor using agricultural policies the primary policy thrust. The policy goal then was making the economy work for the rich and thus create enabling environment for the pursuit of individual benefits. Maximization of personal gain was the end while the means to achieve this goal was the cache of petrodollar amassed from crude oil revenue. An unfortunate situation which informed when the petrodollar was pouring in the country the major beneficiary was capital expenditure in roads, airports, refineries, etc. Not that it is wrong to build your infrastructural base, but prioritization of capital vote over building skills of the people is the crux of the matter. At the least sign of any major stress, Nigeria's vulnerability becomes easily noticeable.

When the oil wealth ran out after every recession, maintaining these infrastructures is a big challenge to government because it will stretch the slim budget to its limits. Against this backdrop, one can say there is a self-induced limit to growth since the country lacks adequate human capital based equipped to address the rhythmic boom and doom oil cycles. Nigeria's predisposition towards infrastructural development is in line with development thinking espoused by the World Bank in those days. Dr Jim Yong Kim in a speech at Indonesia admitted that the Bank was wrong in the emphasis placed on hard infrastructure above human capital investment in the past and that emphasis has now changed to prioritise human development over infrastructural transformation.

The economic reasoning within the policy formulation apparatus never really changed from the World Bank prescription of infrastructural development using revenue from crude oil. This assumption is fundamentally flawed because it failed to take into consideration issue of investing in the people and how it could affect outcome of development. Furthermore, geophysical realities of the country require it saved for a rainy day. A rainy day when the oil windfall winds down or worst still long spikes replicate conditions unfavourable for development. Government's short sightedness will hold waters in a clime where the financial base is built; where through savings the necessary framework for development exist; where commercial banks with excess liquidity are forced to issue loans to the people; where foreign investors on seeing the stable state will allow FDI flow freely into the economy in pursuit of the Petro-dollar; where the people with access to cheap loans engage in productive activities capable of trickling into all sectors. It is also where people with rising income have financial clout to purchase goods and services needed for their families.

This should not be an import dependent country, neither should it be a place where people are fond of using up all their earnings to feed, or purchase goods at credit or worst still, are living on less than $2 a day.Alas, it is the case in Nigeria today. This is a country where the fiscal regime is based on the wrong premise of petroleum revenue continuing *ad infinitum* to fund the budgets. In this economic reasoning, fiscal policy pays insignificant attention to overcoming instability of revenue through appropriate policy. Besides, setting aside revenue through savings in form of foreign reserves or sovereign national fund is not a measure of development in the eyes of undiscerning citizenry.

Development in the eyes of the people is where people are capable of meeting all their needs irrespective of stress in the economy. Since decline in oil revenue is cyclical in nature, there is the imperative for forced saving of excess crude oil fund when oil is in a boom period. Decline in revenue is, therefore, inevitable seeing that lack of reserve after 2014 boom explained excruciating impact of the 2015 oil spikes had on the people. With no flexibility mechanism built into the system, capital expenditure was the first sacrificial victim while the ripple effect is felt across all the sectors of the economy.

One might have understood if capital expenditure brought in additional revenue in form of profit. What we have today is a bottomless pit where over-budgeted and under-funded projects compete for attention. For years the national budget had insignificant attention paid diversification using agriculture as its springboard. Nowhere is the disparity in policy more visible than the amount earmarked for agriculture and capital expenditure. In a country where budgetary allocation favour generating electricity than food production, the country could have energy it needs in the long-run but lacks people to consume energy generated as they are not empowered to use distributed electricity. A primary reason for generating power is to provide households with energy which can be used for developing its agricultural and industrial base. When energy generation is solely a statistical device depicting number of physical counts of projects not real kilowatts of power generated in absolute terms, then, the goal of such action is defeated.

Power in the long-run would be generated though the people to purchase such energy will be in abject poverty. In such situation, the assets might go under for lack of maintenance linked with inadequacy of funds. The task of walking the tight rope between power generation and energy access requires trade-offs. Even in some developed economies it is a difficult problem, these societies have internal mechanisms for addressing the problem. In Nigeria, where the only avenue is the major determinant of all issues it is a recipe for scarcity with scarcity rent starring the poor in the face. Where there are functional institutions, the regulatory apparatus exists to correct ungainly behaviours in energy generation. Nigeria is guilty of chasing diverse issues with one policy tool thus run the risk of confusing policy end with policy means. For instance, different administrations over the years sought to boost agricultural yields through provision of fund using commercial banks to propel such transformation.

To this end, large sums were granted commercial banks for redirection to small holder farmers as loans. What has emerged is the funds were used for short-call loans not to the farmers but were given as loan to big firms. With this development, Nigeria's vision of empowering the poor only served other unintended ends: the creation of a new class of superrich millionaires who benefited from the commercial banks' loans. In this situation, the commercial banks that are presumably the means for delivering policy becomes the end of development thereby marginalizing the smallholder farmers who should be the main end of policy in the first place.

Often, the poor are at receiving end of such policy deficit widening the gap with the rich. Another instance is the food import regime. In present food import regime, the very tool meant to provide solution for reducing the gap in food import and agricultural yields is now the source of inequality. Economically disastrous importation of food will continue with between over $3-$5 billion committed each year to food imports while there is inertia within the productive sector. Understandably, Nigeria followed the prescription of international bureaucracy and their legion of domestic experts. These legions of experts have turned the country into one large scientific laboratory to test pretentiously universalistic concepts not tested anywhere. In doing so have succeeded in fostering universalistic concepts with limited success in other countries where they are implemented yet, flaunted as solution for plethora of problems beleaguering the country.

A country noted to have expressly disregarded IMF advice and still came out of their developmental deficit in Africa is Botswana. According to report, Botswana is one of the highest income countries in Africa today. Although Botswana's per capita annual income was $100 before the independence in 1966 but this has changed. The country like Nigeria is a single commodity dependent economy relying extensively on diamond to grow the economy. According to Joseph Stieglitz, Botswana's success rested on the nation's ability to maintain political consensus, based on a broader sense of national unity. And that political consensus necessary to any workable, social contract between government and the governed had been carefully forge by government, in collaboration with advisers from a variety of public institutions and private foundations. These advisers are the source of Botswana success in mapping out the country's future. Unlike other African countries, the country planned their economy through an extensive process of meetings, seminars, workshops and went to the length of hiring consultants before arriving at any major decision. Interestingly, the country just like Rwanda is blessed with visionary leaders dedicated to extricating their people from the wreckage of a colonial heritage that left them without institutions or human resources. The two countries (Botswana and Nigeria) are emblematic of the contrast that marks the developing world: a contrast between success and failure, between rich and poor, between hope and reality, between what is and what might have been.

Arguably, Nigeria has had her fair share of visionary leaders who have in their own ways imparted positively on the country while other leaders are part of the problem of the country. The widening gap in agricultural food production and population growth has now become a national emergency under the present regime (See Appendix 4 on APP's objectives). With the formulation of APP there is some form of sanctity in the country's import regime with Nigerians now looking inward for local substitutes especially in rice consumption. It may interest that the rot in agricultural sector did not start overnight but is cumulative shortfall of many years of flawed development plans. Coupled with institutional debilities, this made agriculture the weeping child of Nigerian economy.

A solution is increase in budgetary allocation and actions that will bring the Malabo Declaration which Nigeria is a party to into fruition. But increases in capital allocation cannot be justified on the grounds of its fiduciary responsibility; it is the moral responsibility of government to make provision for infrastructural development not at the expense of human development. Key to a sustainable future is to make people the engine of growth. Where there is no human development structural shift towards infrastructural development will make agriculture more dependent on political processes than on human needs.

Policy deficiency in agriculture is, therefore, linked to absence of human capital to efficiently harness resources of the country in building a stable economy. Cultural shift towards capacity building will holistically transform the workforce to be the engine of the growth. There is an undue reliance on capital infrastructure without requisite skill to sustain these projects. A case in point is the Ajaokuta Steel Complex. To all purposes and intent, it is a laudable goal capable of transforming the country, if necessary skills are developed, wherein the industry is in the hands of local workforce committed towards national development. If institutional change through education has been enshrined from the beginning, the decay in Ajaokuta could have been surmount a long time ago.

Of note, education played a pivotal role in the success stories of China, South Korea, Indonesia and India. Cognisant of their weak resource base these emerging economies concentrated on developing their human capital. With skills acquired these countries are adequately equipped to carry out innovative developments. With the rate of human development in India rapid technological innovations have evolved to tackle their energy-related crisis. Note, however, that India is Nigeria's highest trade partner importing crude oil from this country. In a remarkable way India has one of the highest rates of building new solar farms in the world. Generating capacity of the country is growing daily projected to rise to a substantial level in 2030. Evidently, India is planning to extricate itself from enslavement to fossil fuels. Nigeria which is India's major partner should make plans for the gaps this development would create in her export earnings. What has the nation done about this development? Nothing substantial has been carried out. For Nigeria to turn this fiscal deficit into a sizable surplus, millions of youth who are currently unemployed is endowed with necessary skills to return to the land. When the oil wealth dwindles what will keep the country going is not infrastructure but the teeming population young people. Since Nigeria's major source of livelihoods is agriculture, it should be developed in way it can move the productive youth out of poverty while also improving their saving capacity leading to national agricultural renaissance. This, in turn, will encourage the youth to be part of the larger economy of those who take risk. Accepting to participate in risky ventures will promote change not only in orientation but in cultural values.

The trilemma before government today is how to bring about technological, cultural and political shifts in a changing world. To tackle this, the relevant institution must undergo restructuring before we can entrench lasting change. East Asian transformation is the product of gradual and painful change- an item in limited supply in Nigeria. Changes in East Asian societies were not without suffering. Without human capital development opening the country to foreign capital will have only one result: capital flight. When and where capital flight is rampart, the economy is at the lowest ebb thereby causing further depravations.

Refocusing Policy Towards Looming Economic Doldrums

Different strategies are sketch to reduce Nigeria's reliance on fossil fuels. At political and academic circles, assorted solutions are advanced to create a safe passage from the impending economic shocks. Surmounting complexities of embracing a future without oil is still work in progress, however. As part of the strategy, clean energy solutions are deployed concurrently with fossil fuel power generation. The energy sector remains one of the fastest growing sectors and within this category the traditional biomass energy is the dominant sector since it is the primary energy source. A slow pace in imbibing new clean technology is part of the problem. Other issues contributing to the lack of speed in deployment has to do with the low adaptive capacity of the country. For concrete inroad into net zero carbon trajectory, the country should as a matter of urgent need improve the pace of imbibing new technologies especially in bioenergy/biotechnology sector to turn the dream of clean development into reality. The main mechanism for realizing this dream is through integration of traditional biomass sector into new investment schemes capable of offering staggering returns with other co-benefits for host communities.

To fast track the bioenergy sector requires biomass is made the building block for a sustainable energy future. A biomass source capable of fostering immense economic and climate benefits is bamboo. Bamboo will not only ease the negative ecological footprint but will act as a substitute for crude oil. Other sectors are equally important if the country is to win the clean race without its negative effects as follows:

Agriculture—Transformative change requires integrating the largest single employment providers into any plan of action for the future. Agriculture is the dominant employment provider. Agriculture currently accounts for close to 40% of the national income and almost 70% of the working population. While the sector's contribution to GDP will decline in the coming years, its productivity per unit of land, labour, and water will need to increase considerably to feed a rapidly expanding population. Agriculture is a mega business with various classes of people involved in farming activities. The county blessed with large swathes of arable land (Map. 6.1) which should be converted into efficient farms to feed the people.

From the Map above, the abundant agricultural land should be utilized to address demographic pressures. Nigeria's population of 200 m is estimated to increase by 2050 to over 450 m people while biomass consumption will double within the same timeframe. Correspondently, food requirements will be higher requiring improved agricultural yields to feed the teeming population while not placing the fragile environment under further burden. More crucially, food insecurity will be higher requiring

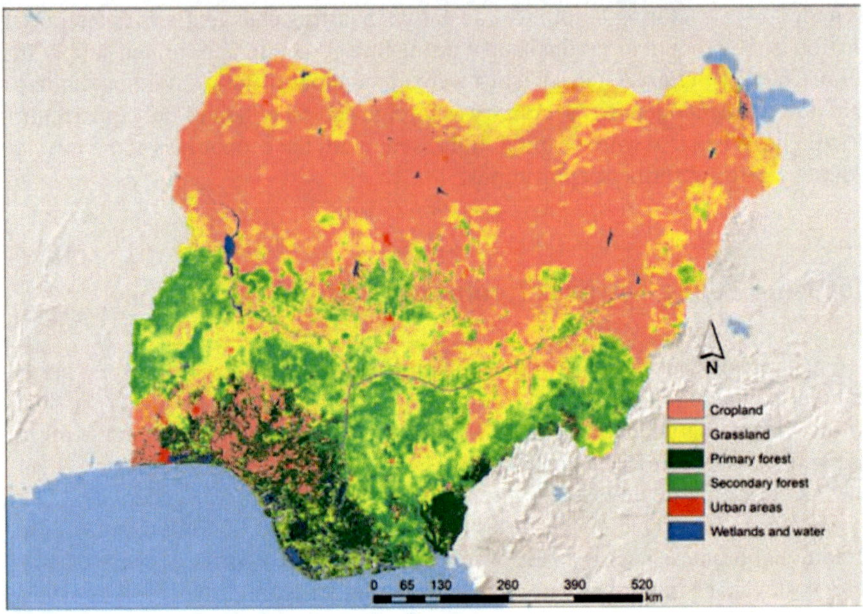

Map. 6.1 Agricultural land use map. *Source* FAO GeoNetwork Database[4]

a carefully articulated food policy. With most farmers struggling to obtain loans to modernize or expand their farms, invest in productive assets or buy inputs, policy levers are needed to drive food production. To address these interrelated issues, government launched the Growth Enhancement Support (GES) scheme in 2012 to supply subsidized inputs to smallholder farmers with over 14 million farmers registered on the scheme in 2015.

More: The Federal Government also launched a series of programmes to increase access to financing, including the Commercial Agricultural Credit Scheme (CACS), the Anchor Borrowers Programme and the Nigeria Incentive-based Risk-sharing System for Agricultural Lending (NIRSAL) (FGN, 2017) as part of the Agriculture Promotion Policy (APP) of 2016. At its fundamental level, the APP seeks to build on ATA of 2011–2015 in working with key stakeholders to build an agribusiness-based economy delivering sustained prosperity through meeting domestic food security goals, generating exports, and supporting sustainable income and job growth. A few specific objectives for the period 2016–2020 have emerged.

Clean Cities and Industries—A green future requires an amiable environment for sustaining industries, building and transport systems of most societies. Large-scale uptake of clean technology could remove waste in society while charting a pathway to efficiency and effectiveness in resource use. Industries embarking on the low carbon trajectory should uphold ecologically sustainable practices, but with an intertemporal equity consideration. For instance, in the transport sector a post-carbon future should emphasize fuel efficiency, stricter standards for passenger and cargo transport, improvement in urban transit, and purchase of hybrid vehicles with high fuel economy. An instance of such transformation is in Lagos State. Lagos state government have championed the drive towards clean city in enshrining clean transport pathways within their economic blueprint.

As a result, clean energy development recorded impressive investments within Lagos state. In other states in the country a few clean energy investments have been recorded. A major factor for this remarkable growth is the active participation of the private sector noted to have been propelling growth in this sector. Notwithstanding, investments in clean energy are inadequate with vast renewable potential largely untapped. Government leadership is vital in channelling investment into clean innovations and uptake of new technologies to tap these vast resources. To jumpstart the process requires active private investments. Small and Medium Enterprises (SMEs) is an important component of rural transformation needed for reordering priorities toward providing critical clean energy services to communities.[5] Because of the vast geographical spread of SMEs, uptake of clean energy would bring about a socially inclusive growth.

Renewables—Since efficiency, productivity and environmental benefits underscore exit from fossil fuels, crucial elements of people's livelihoods must also embrace the energy efficiency changes for a green civilization to take root. Biomass extraction engenders climate risks. For this reason, rebalancing livelihoods of the family farmers will reduce waste and eco-costly inputs. Consequently, should be the stabilization goal of exiting fossil fuel since it can cushion the economic gap a zero-carbon economy entails. Biomass should be integrated into the national development plan with emphasis placed on priorities toward eco-sufficient alternatives in bioenergy production. Climate friendly technologies of solar like the nineteen (19) solar providers granted licenses by the Federal Government should be encouraged. But this should not be mar by handicaps resulting from the stiff ease of doing business.

Power—Power is life essential for sustaining human existence. All human economic activities have energy imperative because exchange of energy underpin modern economic activities. Adequate power or the absence of it determines the nature and form of clean energy technologies deployed and the cost of such green development. The poverty level can be said to be an offshoot of the absence of adequate power supply. While currently a small contributor to energy-related emissions when compared with the oil and gas sector, electricity supply—both grid-based and captive—is likely to experience rapid growth in the coming years as the economy strengthens and energy demand rises with improvement in living standards.

The Electric Power Sector Reform Act passed in 2005 seeks to reposition the sector by changing the structure, and privatizing generation and distribution while retaining transmission under government control. Today, Nigeria has 12.5 GW of installed capacity, but less than one-third is operational (average 3.9 GW in 2015; 3.2 GW in November 2016).[6] Overall, only about 15% of installed capacity is eventually distribute end users, resulting in a huge shortage of electricity supply across the country. Electricity consumption per capita is in rapid decline given the challenges associated with militants' activities within the Niger Delta region. As fallout of this, existing grid-based generation capacity of about 4,052 megawatts (MW) dwindled to only 2,080 MW for a country of over 190 million in 2018. Nevertheless, the price of photo-voltaic panels is falling daily, Nigeria can tap into the vast solar power (see Map. 6.2).

From the map (5.2), it is obvious Nigeria's vast energy potential is underutilized. Historically, the oil and gas sector has been one of the main sources of greenhouse gas (GHG) emissions. Estimated annual emissions in 2010 were approximately 90 million metric tons carbon dioxide equivalent (Mt CO2e) per year, of which the dominant source is gas flaring. The oil and gas exploitation and exploration activities are concentrated in the South and offshore; hydropower in central and southern Nigeria; coal deposits in the South and East; direct solar radiance in the Northeast (orange areas); good photovoltaic (PV) potential is found in most areas; and promising wind sites in the North and offshore.[8]

Increased finance can propel a low carbon growth trajectory since technologies tend to attract higher upfront costs and delayed benefits compared to the higher-carbon technology they can displace. The vast associated gas reserve could also drive growth through increased industrial and domestic use. Lack of adequate infrastructure can limit easy movement of gas from extraction to consumers, and social unrest in the Niger Delta could discourage investment required to upgrade the gas network. Similarly, large swathes of land (see Map 3) offer opportunities for wind, hydropower and geothermal energy generations. Thus far, fossil energy with fossil fuel subsidies is more attractive for the mere fact that it is more stable. To rein in massive fossil dependency, governments can promote mixed sources of energy to tap into the abundant energy potential in the country. The hydro power sector is a promising area where only 7% of the available capacity effectively utilized. To

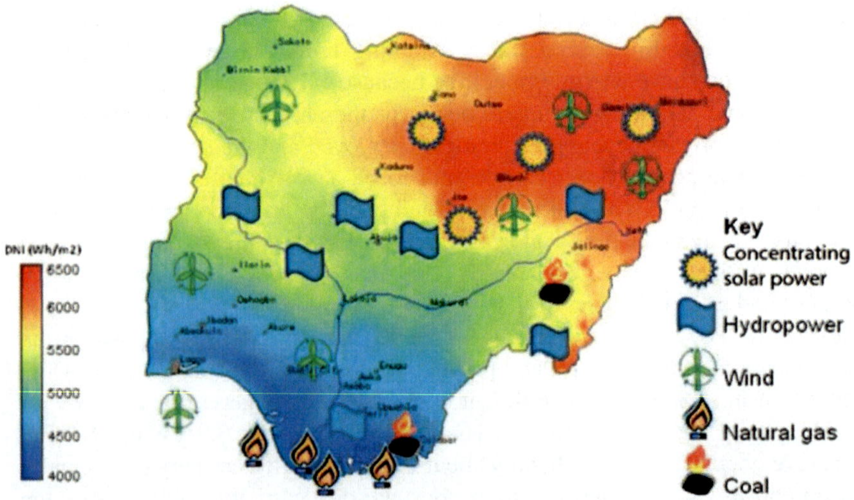

Map. 6.2 Stylized illustration of the distribution of sources of energy. *Source* PVGIS[7]

avert fossil-based growth, government disincentives in the form of eco-tax placed on fossil production and consumption in doing so make it less attractive to energy firms. The envisaged eco-tax should shift the burden from labour to end users and producers creating incentive for businesses to engage in renewable energy. In doing so, force the firms to channel sufficient investments toward carbon free innovations and technologies.

Solid Minerals—A sector largely ignored and is one of Nigeria's most promising growth sectors is the solid mineral sector. Nigeria's vast solid mineral resource has not been tapped and as such made an insignificant contribution to GDP. The contribution to GDP improved since 2015 doubling from N52 billion in 2010 to N103 billion in 2015. In 2015, quarrying and other mining accounted for 89% of the sector, while coal mining represented 7% and metal ores 4%. Most significantly, the country has forty-four (44) known types of minerals of varying mixes and proven quantities concentrated in certain regions. As a matter of fact, every geo-political zone and State has unexplored or under-utilized mineral deposits. Unarguably, the lack of their utilization is due to the attention paid the oil sector and the lack of investment/funding in other viable alternatives to augment revenue base of the country. As a result, mining is rudimentary and predominantly undertaken by small entrepreneurs and unlicensed, unskilled individuals, a situation noted to have aggravated health hazards and environmental degradation.

The solid minerals industry has enormous potential to accelerate the pace of diversification, although constrain by several factors. Crucially, inadequate geosciences information is a barrier to investment. Funding is another matter insufficient to conduct required level of technical work, enforce regulation or complete planned projects. Mining infrastructure, including electricity supply, access roads, and mineral deposit sites is poor. Illegal mining, community unrest and low productivity are drains on the Federal Government's resources whilst the lack of skilled workforce and funding made it difficult to enforce regulations.

A growing amount of policy options to encourage a post-carbon society have emerged. How can we make the most of these options? Can Nigeria really explore its vast renewable opportunities in the post-fossil economy? It is pertinent to call attention to these questions. Efficiency, productivity and environmental benefits underscore the net-zero carbon emission transition goal encapsulate in the NDC. Accordingly, important components of the people's livelihoods must embrace the efficiency regime for a green society to take root.

Rebalancing livelihoods of smallholder farmers to reduce waste and eco-costly extraction should be promoted if we are to realise a carbon neutral future. To end massive fossil production, government must promote mixed sources of energy to tap into the abundant renewable potential of solar, wind and hydro power. The hydro power sector where only 7% of available capacity has been utilized is a promising sector. (The proposed 3005kw hydropower generation in Mambilla Plateau re-awarded in 2018 by the Buhari government is a right step in the right direction. Sadly, it is still entwined in all kinds of controversies with legal tussles between different Chinese and German companies over the actual company awarded the contract a major issue.

As of late, the story filtering out of this sector is not palatable. Surmise to note that corruption has eaten deep into the contract sum escalating the project far beyond its actual value. If such step is link with disincentives in the form of eco-tax and eco-labelling, fossil production and consumption will be less attractive to the fossil fuels industry. The envisaged eco-tax should shift the burden from labour to end users and producers creating incentive for businesses to engage in renewable energy in so doing, channel sufficient investments towards low carbon innovations and technologies.

Development plan should be reoriented to include climate-related emission cuts. To bring this to fruition, domestic laws should clarify how emissions adversely undermine development hence a punishable offence. (Nota bene: the PIGB failed to tackle emissions in the oil producing regions for economic and other unstated reasons). The law must be explicit on applying both incidental and non-incidental emissions' impacts on society. Government failed to incorporate these principles into law, even where they exist its application have been inconsistent amongst various sectors of the economy.

Data on emissions have not been collated consistently, accurately, or on a scale over time relevant to national-level planning on fossil fuel consumption collated with a view to stopping fossil fuels consumption. It is necessary for government to designate a responsible agency or institution with authority to coordinate all issues related NDC by setting up an independent monitoring and evaluation unit with legal backing. The proposed agency should build the national research and data collection capacity to support long-term development of information on monitoring, reporting and verification (MRV) component of the NDCs. Planning for the future should also involve civil society and universities. Capacity to implement the NDC is limited, due to lack of financial and technical resources for attaining the Paris Agreement. The universities and CSOs can fill the void. However, an awareness of the NDC be created in such a manner that incorporates environmental education into the academic programme at all level. The imperative to develop skills-based curriculum in subjects like science, geography, social studies, environmental education and technology that will empower children to better respond to the threats of climate change is high. Retraining teachers on climate change adaptation teaching strategies and techniques should be at pre-primary, primary, secondary and tertiary levels.

Innovative finance is solution for bearing the annual $2 billion cost needed combat climate change shocks. In January 2016, the World Bank Group announced a new plan for $16 billion funding to help countries across Africa adapt to climate change and enhance their resiliency to extreme weather events. The initiative is called the Climate Resilience and Low-Carbon Development as part of the Paris Agreement. Government should avail itself this and other climate funds opportunities available at international level to cushion impact of downward spiral of the economy with the prospects of ending carbon extraction and consumption globally.

Another way out of this impending implosion is to take advantage of the global agencies on electrification to reduce carbon intensive energy generation. There are many agencies committed to Africa's electrification at the international level. Nigeria can take advantage of the large pool of electric energy generation entities or bilateral organizations to bail out of energy-related constrains fostered by the end of Oil Age.

The primary goal of these agencies is to reduce poverty and promote inclusive growth in meeting the basic energy needs of the people. The mechanism to address this need is through efficient utilization of renewable sources such as solar, wind, biomass and hydropower.

These agencies are: Power Africa of the United States Government- committed to provide $7 billion in financial support to add 30,000 megawatts of clean and efficient energy; the New Deal for Energy in Africa of the African Development Bank with the principal aim of universal energy access by 2025; the United Kingdoms' Energy Africa Campaign targeted at achieving universal energy access by 2030; the governments of Italy, UK, USA and Denmark's Sustainable Energy fund for Africa is a $95 million multi donor facility to support small and medium sized projects; EU's Electrification Financing Initiative with a total fund size of $81.8 million and Africa Energy Leaders group of heads of states and corporate executives of companies from Africa.

Addressing the crisis through reliance on power sector financing from multilateral and bilateral organizations alone cannot subsist. To achieve the goal of reducing dangerous anthropogenic climate change necessitates synergic integration of the several components of the NDC with the Vision 20:2020 development policy dovetailed to NASPA and other national priorities such as the new recovery plan for post-COVID-19 era, Sustainable Development Goals and poverty reduction schemes. Government share larger proportion of blame for shortcomings of mismanaging challenges associated with climate change. It has taken limited advantage of the Global Environment Facility. For this reason, proactive action on climate extremes is very low with information on financing risk inadequate. International fund should drive commitments at local level which must be match by the political will to address local challenges. Therefore, the global commitment should reorder priorities toward basic needs of the society with view to limiting poverty and inequalities.

The NASPA document should be update to include financing risk based on region specific challenges integrated into the national framework with a unit committed to monitoring the NDC. The drive toward energy efficient, sustainable and climate friendly economy require more than just reinventing nature as a zero or low carbon economy. Such model will subsist for developed economies where there are internalization and externalization of cost but in Nigeria this approach can reverse narrow gains in environmental progress. Adoption of best practices is vital for energy efficiency and conservation in achieving the national decarbonization objective. Government should implement the Power Sector Reform if significant progress will be made in decarbonisation. Passing the National Climate Change Bill should create high-level political support for the NDC leading to a green society. A green society needs a green bank to efficiently harness resources whiles also empowering the people to face rising challenges in the post-carbon world.

Making the Future Without Oil Energy Efficient

Nigeria's low carbon emission policy is based on business-as-usual approach of responding to anticipated threats using existing practices and management/technical strategies. These are mostly incremental and reactive responses targeted at climate risk without fundamentally engendering systemic changes to challenge existing approach of using incremental responses. For transformational change from fossil-led development to low carbon growth, some of the encumbrances underpinning BAU is resolve to transform the system, structures and values of the people. These changes in systems, values and norms able to incite a low carbon economy are in diverse areas. A change in the roles of political actors and network is the first start point. Political actors and networks grow the economy towards an eco-efficient path. Political actors rising profile in climate mitigation and adaptation hinges on their high political visibility in most communities. For socially inclusive growth these vital community must embrace the gospel of change towards energy efficient alternatives.

Another related issue is building strong foundation for institutional coordination. Nigeria's uninspiring performance in resolving water, energy and food poverty and associated risk of climate change is largely attributable to the existence of weak institutions. This much we have repeatedly said. The country's ambitious goal of diversification can be impaired by weak institutions endowed with ensuring progress in reordering the negative emissions within the country. Inconsistency amongst government agencies and numerous unnecessary conflicts over technical procedures and ownership of projects undermine commitment from these organs of government bestowed with such responsibilities. Often, powerful agencies hijack projects with little or significant requisite supervisory skill. Implementation shortfall in agricultural sector resulted in rolling out of biomass schemes meant to be alternatives to fossil fuels slower than anticipated.

Reversing adverse carbonization together with diversification require a step-change not just in consumption and production of carbon but also institutional transformation. It entails translating political commitment of Paris to concrete steps to build a climate resilient society. A climate resilient society is one where there is a broad based domestic political constitution of support; where government constructs a new political narrative; where overarching goal of development is to implement low carbon policies sufficient to reorder negative emissions of the country. Reversing this trend will task governance structures. A powerful constituency of support needed to roll back Nigeria's carbon profligacy. This is understood as being made up of powerful and evanescent block of captains of industries, Civil Society Organizations, Community Based Organizations and end users. The power dynamics within this odd mix will seriously overturn any potential policies favouring environmental protection.

Though implementation of low-carbon development is gaining traction, it is behind schedule. What we should note here is not just how weak institutions are the major bane of Nigeria's inability to fashion a carbon neutral path to the future. Rather, the ambitious goal of controlling carbon uptake is determined by elite interest at diverse levels ranging from individual, community, local to national scales. Tackling issue of diversification within these scales is generally a political obligation dependent to a large extent on political will which is also dependent on political exigency and how the political class with a short-term horizon view and relate with nature. It also depends on how government improves the risk management capacity in dealing with development difficulties recorded in the country. Dealing with these deficits implies putting on ground new structures to alleviate poverty, promote food security, improve health, risk reduction, education, communication, disaster relief and insurance with these systems interacting in multiple scales with differing effects for sustainable development. When such shift is achieved policies will be based on pro-poor carbon agenda where reducing biophysical alterations, together with socioeconomic degenerations fostered by climate change are ameliorated.

It is important to note that the institutions vested with resolving climate risk through diversification lack capacity to transform the biomass plans to action. Government should strengthen existing institutions such as the NNPC via allocation of financial and administrative resources to foster successful biofuel production whilst also managing risk ensuing from this sector. Leadership most especially from government is a necessary condition for exiting fossil fuels and should be based on the provision of political authority to promote buy-in of all sectors. Replacing age-worn institutions with more adaptive, integrated, multi-level and feasible governance appraisals and inclusive decision-making with the intention of operating successfully across multiple scales will boost adaptive capacity and effectiveness of the diversification response. More importantly, it is the responsibility of government to correct the problem of transversal and vertical coordination, including institutional duplication and weak institutional linkages within all facets of the economy.

Although some states embraced the low-carbon pathway using sustainable development as the roadmap, actions carried out are not really targeted at eliminating fossil dependence. Disconnect exist between developmental goals of the States in agriculture and environmental protection and those of the Federal Government. Given this variance in development goal, dichotomy between sub-national and national goal have magnified existing inequalities. For inclusive green growth national and sub-national objectives for the environment should be based on a unidimensional vision where there is goal congruence between the State and Federal Government. Accordingly, agricultural change will be the fulcrum of post-carbon practices mainstreamed into the national development blueprint.

Since tackling fossil fuel divestment involves actions of all stakeholders who also function as a powerful mobilizing organization in society, these stakeholders when adequately mobilized will pursue the goal of reducing the impact of exit from fossil fuels. This is because local institutions and CBOs are integral parts of the community with their activities based on values and norms of society. Due to the central role these organizations play in communal mitigation and adaptation to climate risk, they would ensure sustainability of agricultural activities within these communities.

Government cannot go it alone. Bilateral and multilateral organizations have roles to play and have over the years played crucial role, immensely contributing to the change in economic and environmental landscape globally. Much has been said about criticality of these organizations to managing the complex environment of economic diversification together with its impact on national development. These organizations provide a holistic lens for understanding and explaining policy processes and how they continue to be influenced by policy changes. They are material to understanding how diversification should encompass more than one complex system at work in the policy processes. This will bring clarity to the complication inherent in a complex system as all the parts interact with each other in a nonlinear way to that keeps changing over time. As these systems are undergoing continuous changes, the CSOs should ensure sustainability of existing policies in anticipation of future shock and stress in the economy. In a complex system where activities of other systems exist independent of the others, the CSO should mobilize stakeholders and policy processes to bring these multifaceted systems together into an efficient system. A reappraisal and implementation of national projects will be achieved through this means. Also, it will refocus national activities towards a low carbon development via identifying agendas, setting solutions and ensuring policies initiated are implemented and properly evaluated.

In this vein, the organised private sector is vital for sustainable and carbon neutral governance. Participation of this sector will guarantee lock-in of government decarbonisation, synergistically integrating the risk sector into all climate actions. Contributions of the private sector to building a low carbon economy could be in the form of Public, Private, Partnership (PPP), Corporate Social Responsibility (CSR) and business model approaches. Various governments have pursued the PPP and CSR approaches with limited attention paid the business model approach. The business model approach will transform the way enterprises do business if the goal of curbing hydrocarbon production is incorporated into their business models. Adopting the business model will realign organizational goal towards disaster risk reduction in the post-fossil Age. Synergistic integration of organizational goal with national development plan will ensure goal congruence between national development plans in the process strengthening PPP critical to enshrining a carbon-free society.

More the fact, establishing linkages between NDC/SDGs are crucial to this objective. The new ERGP is designed in a manner that it is a constituent part of macroeconomic decarbonisation agenda of government. The post COVID design should place strong hope on socially inclusive growth to usher in stable environment for the transition to a low carbon pathway. Whereas in the NDC mitigation has a strong fossil fuel backbone negating any gain of decarbonization. Evidently, exacerbated fossil exploitation cannot fuel low carbon growth. Tensions at multiple fronts on carbon intensive growth and curbing carbon dependency are abundant given that fossil fuel enthusiast are parading all sorts of fossil fuel options to be viable path out of the current recession.

The abundant natural gas is a 'bridge technology' capable of fostering transition to a net negative emission trajectory. The plan to generate energy for all by 2030 in the NEP via large-scale generating plants will not bring about energy sufficiency or climate resiliency. It will accelerate production and consumption of fossil fuel to meet energy demands of the rising population. Nigeria's present decarbonisation activities amount to bailing out some buckets of water as the Titanic sinks, so to say.

Ongoing efforts amount to closing the door on one form of pollution while inviting a more ravenous and even degenerative option through the backdoor. A scale-up of the renewable energy sector should improve energy efficiency while ensuring risk associated with decarbonization is minimized. Operationalization of this objective would task government structures. Hence the political process of low carbon growth gets further alienated from the structures and institution foisting climate solvency.

Mainstreaming specific low carbon growth and other national priorities will facilitate speedy reduction of carbon production. Nevertheless, promoting energy mix involving renewable energy and fossil fuel might receive minuscule attention in the new blueprint and the new INDC submitted after COVIlD-19. Coronavirus will bring out new emergencies requiring further capital outlay. The loans are going to task government priorities making it economically expedient to grow fossil infrastructure on the heels of hydrocarbon production.

Disconnect between different roadmaps targeted at economic growth, diversification and carbon-led growth will create institutional apathy with several arms of government promoting policies with their institutional mandate as its objective. Synchronizing these competing compasses into the national development plan will make it easy for implementation, monitoring and reporting progress in post-COVID era. Equally crucial is the imperative to mainstream the NDC into both the new recovery plan and National Development Plan thus establishing linkages between different roadmaps where technocratic, ecospheric and peoplecentric vistas compete for attention. Effective risk management can reduce harm to the economy posed stopping hydrocarbon extraction in the post-COVID-19 world. A shortfall in accessing/pricing risk at local and national level could be an impediment to energy sovereign future.

A sector with a high propensity of addressing this while also improving resiliency of the vulnerable class is the insurance sector. Though risks associated with agriculture are largely borne at national corporate level, such activities failed to resolve smallholders' vulnerability. This class of farmers lack finance and technical resources to overcome delude of exit from fossil fuel related risk unleashed on them. Even where some forms of insurance have been provided, there is uncertainty about its sustainability. Organized private sector immensely benefited from agriculture-related risk insurance with insignificance insurance for low income earners. Smallholder farmers have inadequate collateral to back cost of insuring farming activities. Government as the last insurer can formulate, implement, and evaluate policies fostering climate risk insurance. These policies should ensure affordable insurance with awareness created within local communities of the benefits of insuring agricultural activities.

Social processes and institutional mechanisms must aim for concerns society value most. Moreover, identification of resilience-enhancing actions such as poverty alleviation scheme can meet both developmental and environmental concerns concurrently. Others are via carbon emission reduction mechanisms such as REDD + with the co-benefits of achieving carbon sequestration while raising additional incomes for the host communities via carbon credit. Clean energy transition is linked effective and flexibility of existing institutions to handle change. Efficient management of climate change entails both public and private sectors utilization of human and non-human nature. The capacity of players within the action arena is critical to managing a complex phenomenon. Viable institutions will incentivise processes to overcome societal pressures, recognising the role traditional knowledge plays in moderating uncertainties. For instance, the importance of indigenous knowledge of floods to Benue communities living along River Benue banks is made the fulcrum of their adaptation strategy.

The country should develop institutional structures effective in resolving the low carbon growth and economic growth dilemma. Such institutional structures is reinforced with a clearly articulated Low Carbon Growth Plan (LCGP) which lays out the direction and vision of development whereby contradictions of growing the economy while also ensuring decarbonization is resolve. The proposed LCGP should define, manage and institutionalize how communities should tackle complex issue of climate change without compromising the country's need for economic growth in the post-COVID era. Drafting of the LCGP will give legal backbone to transition with the LCGP transformational processes linked to a viable institution, strong leadership, organizations and groups with the purpose of promoting green growth.

Developmental needs and climate priorities require policy reforms targeted at reducing all impediments to low carbon drive. Lowering demands for energy through demand side management could lower GHG emission at micro level. For macro level, mitigation of supply side reduction of emission requires enabling law and institutions. Supportive environment for investments in low carbon development should increase private sector participation in policy formulation and implementation. A stable regulatory regime will incentivise decarbonisation fostering a technological shift towards a low carbon growth. Such incentives include: subsidy reform, carbon taxes, eco-labelling, Emission Trading Systems (ETs) and green bonds will send long-term policy signals to investors while long-term planning will help establish integrated adaptation providing financing and access to technology.

Paris agreement and now the coronavirus pandemic are credited with creating renewed momentum for a robust carbon market instrument which will enable the country meet most of its obligations. Carbon pricing initiative including development of ETS will increase Nigeria's potentials to realise the decarbonisation goal through provision of financial support for clean energy transition. Technology is the key to opening a fossil free future; deployment of innovative technologies would catalyse green growth in so doing avoid the eco-costly and inefficient technologies of the past. To reduce the widening technology gap, investments should be scaled-up in renewable technology. To catalyse investments, an environment favourable to risk-returns is create to stem negative effects of investors' apathy resulting from low rate of capital returns. With appropriate regulatory framework advanced clean technologies should be deployed to support projects with higher climate change benefits.

Capacity building to cope with the requirements of low carbon technology will adequately equip the workforce to manage the complex environment of climate change. Training the next generation of high skilled workers who will become entry level technicians, engineers, managers and entrepreneurs would enable the country meet the climate target. For now, Nigeria's renewable energy drive needs policy levers to catalyse transformative change. Transforming these resources require international support from various financing mechanisms and leveraging Green Climate Fund, Adaptation Fund, Reducing Emissions from Deforestation and Forest Degradation (REDD+), Forest Carbon Partnership Facility (FCPF), National Appropriate Mitigation Actions (NAMAs) to reduce fiscal pressures and attract private sector participation in sustaining the low carbon development.

Generating funds within the country is vital in financing green initiatives. The creation of National Climate Fund and issuance of green bonds will encourage eco-friendly development. It could also justify donor participation in clean energy projects locally thus bringing synergy between public, private and donor community on financing economic diversification. A favourable environment for investment would reduce risk, build capacity, promote knowledge transfer, and create awareness of low carbon growth. Overcoming these uncertainties entail credit-risk guarantee.

Proactive steps on synergising decarbonisation to the Sustainable Development agenda will be needed in the years to come. This can be achieved when the country makes good governance bedrock of its post-Paris and post COVID-19 climate risk reduction. To make decarbonisation work for the poor, a stable policy arena and effective leadership would provide the drive for change. Intensifying scale-up of existing efforts guarantees eco-efficiency thus strengthening people's capacities to cope with emergent risk such the coronavirus disease. Designing policies and operations to optimize multiple benefits from actions targeted at Paris Agreement, SDGs, the Sendai Framework and the new recovery plan/national vision is crucial to Nigeria's low carbon growth. For this to happen local and national policies strategies should be inclusive of approaches for optimizing multiple benefits.

Different end-use technologies with multiple benefits are being deployed globally to tackle the climate emergency. The policy arena in the country should change in line with these technological improvements. Government leadership is pivotal to effective utilization of resources with the spill over effect of policies resulting in pollution cutback. Government's role of providing basic needs of society is heavily taxed in the advent of the economic recession. Pursuit of developmental requirements should, therefore, aim at tackling climate risk which is closely intertwined with other issues. Moreover, climate change is the outcome of externality involving inequitable and undervalued allocation of resources. It is equally linked with poverty and inequalities in society hence redistribution and reallocation of resources between all facets of the economy which is important in ameliorating climate change.

A particularly crucial role government should play includes addressing international commitments on mitigation and adaptation to climate change by building a low-carbon and climate resilient path out of the current existential threat. Nigeria aims at reducing emission intensity using inclusive policies that improve human well-being. Creating an enabling environment for implementation of the NDC demands new fiscal policy measures or instruments. Policy levers put in place can propel national process towards an equitable carbon regime. A comprehensive overhaul of the fiscal policy linking socio-economic outcomes of loss and damage to elimination of carbon will change the emission intensity of Nigeria's GDP. This goal will require sector specific actions targeted at lowering demand for energy using demand side management which should reduce carbon intensity of the economy. Other measures include cap and trade, removal of subsidy, appliance rating, eco-labelling and eco-tax. Fiscal incentives to promote renewables including subsidy, tax holiday, investment grants and import exemption will incentivise prospective investors to be compliant with national decarbonisation goal.

The scale of mobilization needed to catalyse a fossil fuel exit requires a pro-agriculture agenda. This would have need of a carbon tax imposed over firms with high carbon footprint as a tool of redistributing income and reordering negative emissions of the country. Likewise, within urban centres building codes for green cities be introduce to lessen GHG emissions. As a measure to address climate change such measures are suitable to the country while improving energy generating capacity of the plants. But the implementation of these policies is proving inadequate.

All said, there is no single acceptable path to sustaining environmental protection related challenges now and in the future. As coronavirus is still ravaging the global economy, addressing multiple scale stressors such as climate risk will be a burden to an overburdened hydrocarbon production hence will induce diversification. But it involves finding an equitable strategy that achieves social, economic and environmental goals concurrently. The health infrastructure is already overwhelmed by the pandemic. Resolving these miasmas of risks entail varied transformation within society and this is hinge on context, policies and participation of stakeholders, amongst other issues. Addressing the issue calls for trade-off between competing choices, needs and goals of society. These goals are changing daily with new health reality of the country beating the tune.

Towards an Inclusive Green Growth Society

Can a new future emerge from a world enmeshed in new health and economic uncertainties? Will Nigeria jettison her growth for environmental quality when the people are hungry? With these perplexing questions we open the discussion on building an inclusive green growth society. There is divergence in views on tackling the carbon upload and economic growth while also ensuring health security in post-COVID-19 age. Divergence abounds between diametrically opposed agenda pursued in realising GHG emission reduction objective of government and building a stable economy in a coronavirus infested world. The disjuncture is indicative of a wider international level divergence railroading environmental protection goals into other national agenda. Pursuing other agenda is indicative of a wider developmental deficit recorded in the wake of COVID-19 requiring us realign priorities towards developmental needs as uncertainties ensuing from the international oil price volatility mounts.

Reducing greenhouse gases is already forced on the country not by choice with low market resulting from lockdowns and need for social distancing. Even if the GHG increase, reducing emission is definitely not a day's job for the mere fact the 'iron law of climate policy' (to use Pielke's term) subsists in most communities. This law assumes in the tussle between environmental progress and economic growth the latter always wins. The assumption here is some Nigerians will prefer short-term economic growth gains than pander to long-term environmental protection needs.

The reason is simple: improved environmental quality is only attainable within the next generation's life span arguably. Thus, attaining this goal is not of high priority for the elite with a myopic vision of our economic life. So, it may seem Nigerian elite vested with maintaining the fragile ecological equilibrium also hold a Cornucopian view of technology where the 'Iron Law of Climate Policy' rules over all developmental activities. Nigeria's deployment of technology only explicates elites abiding faith in technological fixes capacity to meet challenges of initiating near-term material well-being for the people- An explanation for the immediate goal of the ERGP (which expires this year) seeking to promote fossil fuel production and consumption using technology possibly avoiding eco-costly domestic environmental challenges.

Nigeria's approach to building a virile society is largely one of a short-term reactive approach to impetuses from the international bureaucracy. Fundamentally, they are incremental tweaks at the problem but will not get its foot on the door of climate insolvency. Short-term reactive actions cannot suffice to extricate the country from strangulating hold of endemic energy, water and food poverty ravaging the people. Also, short-term measures could hurt the poor due to their extensive reliance on biomass for domestic energy purposes. In this reasoning, pollution is necessary evil needed to sustain the smallholder's family. With growth and pollution coupled both at micro and macro levels, the end is not at sight but rather will reinvigorate fledging wheels of fossil fuel consumption.

Resolving the antithetical contradiction of cohabiting opposites and unholy matrimony of growth and pollution require structural shifts. Envisioned shift needed should be an all-encompassing systemic change such that there are linkages between economic development and environmental protection, and the new decadal national vision, post-COVID-19 recovery plan and resubmission of INDC for 2020. This can be such that there are pro-poor measures linked pro-market approach to bring synergy that catalyse inclusive growth. In this new era, government's actions are targeted at transforming society at all scales with the spill over effects of such policies engendering environmental quality. Such activities are the added benefits of policies but not the goal of development.

Since there is discrepancy in distribution of income fostered by coronavirus, equitable allocation of resources is a double imperative if the decarbonisation drive should succeed. This is an outcrop of its centrality to the NDC and poverty alleviation schemes of government to ameliorate human health-related faced by the people. A bottom-up redistribution of income will endogenize income distribution increase, GHG removal capacity of the country while also ensuring their resiliency to withstand any epidemic. With the poor pushed out of their comfort zone by poverty, income stability will reduce adverse activities such as depletion of forest resources hence should improve the forest, tree cover and increase its carbon sequestration. Lowering demands for energy via demand side management add value-added activities capable of reducing the carbon intensity of the country. Pursuit of this goal requires sector specific activities targeted at end-users who are contributing to the high carbon uptake of the economy. A low carbon growth for a technologically handicapped country entails depletion of natural resources with some form of eco-efficiency thrown in. Depletion to a large extent is a major driver of growth leading to an inevitable end: negative growth. It then means there must be a balance such that there is an equilibrium between the competing needs before growth can pick up in the real sense. Growth, here, for the economically handicapped country will involve some level of depletion. What Nigeria needs in the coronavirus age is not a total curtailment of resource exploration, but sustainable utilization for development purpose to end endemic poverty and raise foreign exchange to repay her debt.

To recap, different forms of transformations are happening, but they have not, so far, triggered a low-carbon revolution or fostered an inclusive green growth path. COVID-19 short-circuited government's goal for the people and the environment. Though the environment is a major gainer in short-term, the environment is a long-term loser since hydrocarbon deposits will fuel growth. Though government initiated a few structural changes targeted at promoting a path to a low-carbon growth, these actions are aimed at poverty alleviation and socioeconomic development with the expectation of meeting developmental needs of the people. Here, the primary benefit of carrying out socioeconomic development is not environmental protection; instead low-carbon growth is a secondary benefit. Climate actions are primarily to incentivise society towards a higher standard of living, but not directly targeted at promoting low-carbon growth.

But this should not be. Policy lever can reverse the negative trend. Some policy levers to sustain low carbon growth in post COVID-19 era should foster a dynamic optimization approach capable of maximizing linkages between the public and private sectors, if a low carbon inclusive growth must take root. These measures should maximize private utilization of resources but should be premised on the availability of natural resources while improving human well-being. In many respects, the transition to a clean energy future warrants optimising agricultural production simultaneously with promoting renewables to replace fossil fuels. As the country continues to grapple with an exit formula, the current funding of capital expenditure indicates an alarming lack of regard for tangible to-do list to end fossil fuel dependence. There are no clear-cut lines specifying modalities for reducing fossil fuel power generation. Along these lines, it is plain the fossil-led economy is gaining footing when compared to renewables. Probably the most visible consequence of the current intervention in agriculture is the Anchor Borrowers' Programme. Likewise, the FADAMA I and II have mutated to FADAMA II(I)–Additional Financing (AF) with additional N8.3 billion loan from the World Bank meant for 5634 graduates trained in agriculture value chain to be gainfully employed.

The scale of the FADAMA projects will improve rice and tomatoes yields thus reduce the manifold pressures on the foreign exchange from this sector. Further, the youths engaged in the programme would become part of a vibrant and promising transformation of the agricultural sector. In that sense, the agricultural sector would be revolutionized since agriculture ideas based on business plans verified and approved by experts can promote agricultural development. There is additional concern in respect of renewables' role in both the energy-mix and economic diversification initiative in post-carbon/COVID world. Given the minimal progress made towards meeting the Nigerian renewable energy masterplan, Vision 30:3030 and other renewable energy-related schemes, it is clear these targets are not realizable within the confines of the prevailing socioeconomic environment of the country. This fate has befallen the Vision 20:2020 which was derailed by several socioeconomic and political factors.

Similarly, the Vision 30:3030 will meet same fate as Vision 20:2020. For one thing, the renewable energy uptake is beset with a menu of challenges hence it has remained small and relatively slow-growing on the balance. The facts remain, however, that low carbon technology is material to disentangling from fossil fuels. So, one can say the sector crucial to the post-fossil future is evolving but is stifle by policy inconsistencies causing serious uncertainties within the sector. No doubt, Nigeria is in the position of reaping colossal benefits through scale-up of the renewable sector. Solid investments in the sector can catalyse growth moving the economy away from hydrocarbon dependence which is plunging the country into to the present economic downturn.

Clearly, the world is moving away from the carbon-intensive pathway to a new era- this age is the post-Carbon Age. Based on this, one can say that the sixth wave of industrial growth has commenced. In this eco-historical moment where COVID-19 pandemic is ravaging many communities, fossil fuels consumption which was becoming a relic of the past hence destined for the backwoods of history is now raising its ugly. The reinvention fossil fuel is out of economic exigency. It is part of the ongoing rebirth of the Anthropocene Age which is focused on decoupling growth from natural material use although it is with uncertainties fostered by our current unsustainable consumption of non-human nature. The Anthropocene Age is not all about ending fossil fuels but is a new lens on how society looks at growth through a cross-critical local and global environmental threshold to move away from BAU 'brown' resource intensive pathways to the post-carbon society. In this age, we accept uncertainties as the new normal. This requires new tools for addressing an uncertain world. A new path capable of addressing this need concurrently ensuring economic viability of a business entity is known as the circular economy. An appropriate resource capable of sustaining a circular economy is bamboo.

Bamboo: Tool for Building a Post-Fossil Society?

Bamboo is one amongst several nature-based solutions to tackling climate change. It is a fast-growing woody grass grown mostly in the tropics and subtropics in mixed forests or as pure stands cultivated in plantations, on homesteads and on farms.[4] They are grown for their long, usually hollow, stems (called culms) used or sectioned poles yielding softwood and fibre for processing. Bamboo is an economic grass with both economic and ecological benefits. The ecological benefit of bamboos is largely a function of the carbon footprint together with the sink and stock function. As a carbon stock, bamboos play immense role in reforestation and afforestation programs and are critical for reducing deforestation while as sink its carbon sequestration capacity enables a stable carbon balance in the ecosystem. Bamboos provide millions of people all over the world the source of their livelihoods.

One of the major attractions of bamboo is the ability to grow faster than most trees; growing at up to a meter per day.[10] It is a grass highly renewable with regenerative capacity to maintain the ecological equilibrium of an ecosystem. Another advantage of bamboo is from the extensive rhizome and root systems which can extend up to 100 km/ha and live for a hundred years. Unlike trees the regular and selective harvesting of bamboo culms will not eliminate the plant nor has any adverse effect on the ecosystem, and below-ground carbon in the soil and rhizome is not emit because bamboo forests continue to live after harvest. Though culms that emerge from the rhizomes die naturally after about 10 years if not harvested before.[11] INBAR Working Paper 65 noted how rhizome system survive harvesting of individual culms, so the bamboo ecosystem can still be productive whilst continuing to store carbon since new culms will replace the harvested ones.[12]

Despite this, cultivation of bamboo has not received the attention it deserves within academic and political circles. In Nigeria, bamboos' vast potentials have been underutilized with the importance limited to scaffolding in buildings, fencing, yam stakes and handcrafts. At present bamboos production is uncoordinated with various communities growing the plant without taking advantage of its significance for increasing foreign exchange earnings when exported. Nigeria's bamboo estate is being depleted at an alarming rate just like its forest cover. According to the Raw Materials Research and Development Council (RMRDC), Nigeria's bamboo estate stood at 10% of the country's land area in 1996 but today is less than 6%. The RMRDC further reported that the total useable volume of wood down to 30 cm cutting diameter in the reserved forest areas of Nigeria was 293,775,500m3.[13] A study conducted by Ogunwusi and Onwualu, describes bamboo availability in the country as occasional because less than 3.0–5.9% of the natural vegetation is occupied by it.[14] And that the most endowed States in terms of bamboo occurrence are Ogun, Oyo, Osun, Ondo, Edo, Delta, Rivers, Akwa-Ibom, Cross-River, Abia, Ebonyi, Enugu, Anambra and Imo States. The duo found that in these states at least 10% of the natural vegetation is bamboo dominated with existing clumps showing appreciable gregarious growth contiguous over large areas. Furthermore, Lagos, Ekiti, Bayelsa, Kogi, Kwara, Benue and Nassarawa States belong to the second category where bamboo distribution is frequent with at least 6.0–9.0% of the natural vegetation is bamboo dominated. Pockets of bamboo clumps found in Niger, Taraba and Plateau states as well as within the Federal Capital Territory.[15]

Different instruments and diverse solution have evolved to reduce the adverse effects of climate change. At different points in the geological history of the world different solutions to reducing the build-up of GHG have been advanced, but with limited success. One of the approaches to tackling climate change at the international level gaining traction lately is bamboo. Bamboos' role in the mitigation and adaptation to climate risk is recognised at international level. A total of 50 countries keyed into the global decarbonisation initiative using bamboos to rebalance their atmospheric carbon budget. Nigeria is one of the International Bamboo and Rattan (INBAR) organization member states. Despite membership of INBAR, the country does not have any specific law guiding cultivation and utilization of bamboos. It is pertinent to note that the national policy on forest resources is meant to capture bamboo's development. As such, it is not certain what bamboo's real economic importance is. Though membership of INBAR can potentially drive change towards sensitization of the public on the beneficial attributes of bamboos but cannot catalyse radical change in cultivation. Ultimately, Nigeria's goal of joining INBAR is to explore the country's vast bamboo potentials to create wealth. Sorry to say, institutional challenges have stifled realization of this objective. Even though bamboo cultivation is part of a larger policy on biomass sector but no institutional structure or organization is committed to exploring bamboos in meeting the NDC targets. One would expect that there should be organizations with an integrated programme on bamboo cultivation with the goal of optimally exploiting the vast economic benefits.

A systemic approach to bamboo production whereby set targets for its cultivation are implemented to guide production is non-existent. Lacking any institutional standardization and coordination bamboo potentials remained not only largely untapped but the most preferred species such as *Khaya ivorensis, Daniella oliveri*, etc., has become scarce in the forest reserves and the natural forests.[16] The result is the growing of bamboos by smallholder farmers who lack both technical and financial resources to make bamboo cultivation a viable venture.

It is against this backdrop of widespread cultivation of bamboos in Nigeria that we promote its role in the post-Oil Age. Bamboo is an economic grass capable of ushering an era where consumption and production of forest resources are ecologically neutral. Consequently, cultivation of bamboo is catching on in some countries. Rising interest in bamboo promises to be one of the major transformations in developing economies due to its economic value. Bamboos intervention could result in a transformational change not only in how the country makes use of biological materials, but in ensuring equitable use of ecological assets. It promises to be for the twenty-first century what palm tree was to the twentieth century national economy. With the rising profile of bamboo, it will replace other forest resources used in providing the economically active population with their energy requirements as well as a new source of livelihood.

Within the agricultural sector, the use of agricultural materials as an alternate source of energy is not entirely new. Bamboo immense benefits have been tapped by different communities in the past; but it is in the past decade that the importance to maintaining a stable carbon budget for the global economy become more evident. In the last decade bamboos are gradually becoming a valuable resource. Bamboos' mercurial rise to global consciousness is outcrop of the crucial role it plays in stabilizing the ecosystem, repairing soil ruined from degradation and reforesting land deforested through economic activities. Increases in interest in bamboos and its rise to global eminence is largely attributable to the pivotal role INBAR played in the dissemination of information and resolving technical issues hitherto impairing its cultivation. INBAR in conjunction with different governments (most especially the Chinese government) promoted bamboo cultivation globally. The level of efficiency bamboo promotes in the drive to reduce high carbon uptake of the global economy is the main selling point. For sure, bamboos ensure energy efficiency in a manner not adding to the pollution burden of the overburdened global economy. Even when bamboos are harvested for domestic and economic activities, the physical stock remains unchanged due to its fast maturing and regenerative potentials.[17] Based on this, bamboo products should be the kernel of green growth since large swathes of land can be restore through its regenerative capacity.

Bamboos effectiveness in transforming Nigeria's consumption of firewood is not in doubt. What is in doubt is how the economic class avails themselves with its vast opportunities. Overcoming these encumbrances hinge on the country's goal for the ecosystem. In a country where carbon consumption causing climate change is high, where fossil fuel drives growth, policies formulated by the ruling class for political expediency must be slanted towards the dominant sector. Here, ecological imperatives are not the guiding thread of policy instead it is political fortunes of the ruling elite. Even though bamboo can foster climate resiliency through its carbon sequestration function and carbon credit potentials, it would be relegated to the backwoods of history. The co-benefits of bamboos are from its tendency to improve environmental quality, health; economic wellbeing in the communities such that bamboo is the engine of energy consumption.

Bamboo is not visible in the policy arena with insignificant attention paid its cultivation. With suitable environment in the South-South zone, the country can exploit its bamboo to restore land degraded through pollution from oil exploration. More the fact, bamboos are highly resilient and can grow in most conditions. If bamboo cultivation is properly conceived, it will promote non-fossilized biological materials as feedstock in the production and consumption of energy. The imperative to integrate a sector of the economy with both economic and ecological values to the rural communities is high. Integration of bamboo into the national development plan will not only limit adverse carbonization of the biosphere, but agriculture will be given further boost and made the bedrock of national development.

Limitations of Bamboos to Strengthening the Low Carbon Growth

Various issues have impaired using bamboos to strengthen Nigeria's low carbon development. Prominent amongst the constraints are the want of large-scale cultivation of bamboo and formulation of policy on using the plant to catalyse low carbon growth in Nigeria. Other constraint is the definition of 'bamboo forest'. The definition of bamboo is inconsistent and incompatible across different countries, complicating the assessment of existing bamboo resources and bamboos' potential contribution to mitigating climate change.[12] This lack of definitional clarity at international level created indifference within varied countries on exploring its immense benefits. For instance, the argument within government circle is the classification of bamboos as either grass or trees might reduce or increase its carbon footprint. Lack of standardization of carbon accounting within the various international institutions is an impediment to bamboo's cultivation.

Another related challenge is dearth of accurate data on the inventory of bamboos reflected in inadequacy of determining bamboo cultivation in the country. The ability to accurately paint a concrete picture of the distribution of bamboos and the various species grown reduced its capacity to offset the carbon deficit. The want for publicly available and transparent data will undermine commitment towards using bamboos in reducing their carbon footprint. A limited number of Nigerian farmers have cultivated fast maturing bamboos and are reaping immense gains. Within these communities, cultivation of bamboo is largely in isolated farms lacking any systematic approach. In these communities bamboo is perceived as wild specie with insignificant need for proper management. Furthermore, there is the phobia within some farming communities of bamboo farms harbouring snakes making its cultivation not widespread. Moreover, knowledge of existing species of bamboos critical to determining its carbon intensity has not helped commitment towards its cultivation. At present, Nigeria has not mapped its bamboos spread, the species grown in the country and economic value in systematic way. A proper study of the species will plot the distribution of bamboos such that the non-profitable and low yielding species will be deemphasized.

A critical handicap to cultivation and utilization of bamboo is institutional hiccup which is continually impeding uptake of clean technology within the bamboo producing area. To effectively utilize its bye-products, a new institution vested with promoting bamboos utilization backed by law should be mainstreamed into the national development plan. Creation of new institution is vital when considering using technology for bamboo end products and its cultivation at a large scale if it is to bring about a switch low carbon development. The problem is not so much because technologies are expensive (in fact, they are cheap) but with low adaptive skill the country cannot optimally utilise these technologies.

Institutional inertia is closely interrelated with the lack of political will. Maximization of bamboos should be the driving force of most policies formulated on low carbon growth. But the optimisation of gains from bamboo cultivation faces impediments from a strong fossil fuel lobby with a strong hold on government. This economically self interested group have stymied implementation of an appropriate building block for a green society to take root. To address this, Nigeria should realign institutional incentives to facilitate achievement of the goal of economic development simultaneously bringing about equitable consumption of natural resources. Political commitment and support of government should give policy direction for the development of regulatory instruments for the advancement of clean energy sources such as bamboos. Conducive policy instruments stimulating development of bamboos are important components of the green economy. For effective transition, all stakeholders must be carried along to attain a socially inclusive green growth. Even then, policy inconsistencies persistently weaken attainment of the clean energy objectives of government.

Government should play a pivotal role in ensuring bamboos are made an integral part of the country's decarbonisation. It is unfortunate the country has not committed adequate funds to foster the transition from hardwood such as rosewood (*Pterocarpus erinaceus*) depletion to bamboo as an alternative. New sources of funds and resources are important components of implementing institutional change. Paucity of funds will stretch government pulling resources away from other vital developmental goals. Foreign assistance in the form of carbon credit earned from bamboo cultivation will give Nigeria the financial clout for meeting commitment to sustaining the environment.

Imperative for a National Policy on Bamboo Cultivation

The largest fossil-based markets and car manufacturing economies have announced plans to phase out fossil fuel-burning cars by 2025 and 2030. The move away from fossil fuels acts as a trigger towards low carbon technologies since the prices of wind and solar power are also falling. The reduced cost of EVs means its sales will outstrip traditional Petro-powered cars in the near future signalling economic turmoil for fossil dependent economies. This much we have stated in other parts of this study. What is new here is what Nigeria can do with bamboos to reduce her carbon upload while also increasing the revenue base of the economy in the advent of the decarbonisation of the global economy through reduction of fossil fuel share of the energy mix. Although Nigeria has not changed in line with the global trend towards a low carbon alternative but hedged her bet on fossil fuel being capable of fostering epochal shift in the exploitation of natural resources simultaneously acting as the engine of growth. With the Fossil Fuels Exit Strategy receiving widespread attention globally, fashioning a viable path towards exit from oil is inevitable if the country must reap the huge benefits this venture presents.

Formulating a national plan of action on bamboo serves this purpose if it can reduce uncertainties associated with reprioritising consumption and production of hydrocarbon. This blueprint should speak to what the country's power generation mix should be and the share of bamboos in the energy mix; in doing so, it would mainstream bamboo to other development goals to become the exit from fossil fuel strategy. Finding alternatives to fossil fuel implies a huge gap is unfolding warranting discovering a readily available substitute which should be made fulcrum of Nigeria's low carbon growth. Nevertheless, a miasma of socioeconomic constraints plagues the Paris Agreement making the global goal of downscaling GHG uptake an undecided event. At the heart of this uncertainty is the objective of limiting carbonisation to 1.5 degrees while also achieving zero hunger in 2030. These two goals are targeted at tackling the crosscutting maladies of hunger and climate change thus opening new ray of hope for the world to end two challenges of global poverty and fossil-led emissions. Nigeria's post-Kyoto and food secured society can be made bright if the country can remove uncertainties surrounding the exploitation of natural resources and burning of biomass fuel. Collectively, these two issues have given the country an adverse carbon footprint. A policy framework on bamboo and rattan will meet Nigeria's dual goal of setting new priorities for climate governance based on nationally determined goals and responsibilities which should bring about zero hunger in the country.

Carbon credit is part of the Clean Development Mechanism (CDM) negotiated at Kyoto. A carbon credit is equivalent to the emission of a tonne of carbon dioxide. Compared to the current rate fixed in the Kyoto Protocol, it allows the holder emit more gases. Carbon credits are allowed states or companies participating in the GHG emissions reduction. Such carbon sequestration activity is with the purpose of helping participating countries to respect their commitments to the Kyoto Protocol and now the Paris Agreement. Generally, funds used for carbon credit are mostly derived from the $130 billion-dollar GEF set aside by rich countries as part of their commitment to the Kyoto Protocol. It is based on the idea the best way to control climate change is to transform carbon emissions into exchangeable goods on the international market. A good example of where this scheme featured prominently in Nigeria is Cross Rivers State. Bamboos can have significant contribution to all aspects of REDD + projects, *inter alia* reducing emissions from deforestation, reducing emissions from forest degradation, conservation of forest carbon stocks, sustainable management of forests and enhancement of forest carbon stocks.

In the aftermath of the Paris Agreement, many investment portfolios are now in search of new opportunities to invest in alternative energy globally. The quest to invest in new enterprises lies with the fact that existing investment portfolios most especially in fossil fuel are risky investment thus are unsustainable. Investing in bamboo is a safe and profitable investment in an unsteady financial context owing to the fast-maturing and regenerative tendencies. This explains why bamboo considered in the realm of traditional agriculture moved up the agricultural value chain globally. The recent traction of bamboo hinges on its ecological attribute of reducing the GHG balance in this manner increasing the downstream risks to fossil producers and consumers. Ecological benefits of bamboos have increased its attractiveness to investors who are targeting agricultural production to gain immediate returns on investments.

What is more, in the country's climate is changing daily rising in proportion with the emission of greenhouse gases. Now more than ever in the history of the country, extreme weather events such as floods, soil erosion, heat waves, drought, rising sea level and desertification have devastating effects on the people.

Cognizant of the country's vulnerability the Buhari Administration is seeking new avenues to reduce reliance on fossil fuel. As already noted, one of the alternative energies with immense potential of guaranteeing energy security is bamboo. Production of bioenergy from bamboos is a radical move towards decarbonisation of the biosphere while also providing additional energy for host communities. The radical change triggered by the interest in bamboos pose no serious threat to food crops like maize, soybeans, sugar cane, cassava, palm oil and a host of other food crops used as feedstock in biofuel energy generation. Its high potential to overcome food crop and feedstock complex in the agricultural value chain is the main selling point. Large scale cultivation of bamboo's signal achievement will be as an alternative to using food crops for bioenergy.

More significantly, the issue of food security is vital to government's agricultural transformative agenda. Threats to food security underpin bamboo's importance to national development in China. A strong incentive to invest in bamboo in China is its prospective of guaranteeing food for all thus overcoming food production related problems causing low agricultural yields. Another reason for bamboo's rising profile is the intrinsic tendency to bring about dietary changes resulting from the shift to a rich food product. With food shortages experienced in different countries where there is water scarcity, it is very important Nigeria takes advantage of the cultivation of bamboos to reduce food shortages while also earning additional income. To avoid committing the mistake of unduly depending on the importation of food item such as rice and wheat, cultivation of bamboos can guarantee access to food during periods when there are food shortages. In this way, the country will not only reduce their import bills but will improve their food security.

Global warming is accelerating crop failure presenting new complications to the food scarcity crisis. Constraints to food production are not impaired by global warming alone because rising oil prices are known to affect food prices together with COVID pandemic are causing inflation. With the general state of health-related insecurity, rising oil price will threaten food supply in the country. Demand for bamboo to be use as food item will be further exacerbated by demographic pressure placing severe strains on the food system. Population growth and urbanisation are near-term threats to food supply changing dietary and consumption patterns globally. It is, therefore, a strategic choice to make bamboo a fast maturing grass with high regenerative capacity an alternative food source. Moreover, population growth comes with its own ills in the form of large-scale degradation causing serious threats to the fragile ecosystem. Forest resources provide income to many Nigerians but this economic activity is an ecological negative when deforestation resulting from timber exploitation for large-scale export depletes the ecosystem rendering it incapable of providing support to the planet. Given that Nigeria's natural resources are vital for maintaining the ecological equilibrium but depletion will disrupt ecosystem regenerative activities in doing so poses serious threat to the life support systems.

Bamboos for Sustainable Development

Bamboo cultivation is one of the old guards of the biomass economy vital for reducing demographic pressures. With the country's population projected to be over 412 m by 2050 consumption of forest resources will take centre stage. Meeting growing needs for energy and other requirements of life will increase with the population explosion. Rising demands for forest resources means that the fragile ecosystem will be stretched beyond its capacity. In order not travel on this torturous road, bamboos offer new opportunities to overcome this challenge with additional gains from trade in carbon credit thus engendering a healthy environment capable of meeting both human and non-human nature's regenerative requirements. Overcoming these difficulties will hinge on what government values most: short-term gains or ecological wellbeing of the ecosystem. At the international level, INBAR is championing the drive for optimization of bamboos in the standardization of carbon intensity of the grass. Standardization will add value to bamboo thereby changing people's perception of it both as an economic and ecological plant. Standardization of the methodology to access bamboos' carbon footprint will ensure communities diversify their economies, improve health, access clean water, and healthy environments (through reforestation, preservation, and sanitation) in the process will break the cycles of poverty; allowing a healthier and more sustainable communities to emerge.

Nevertheless, the cultivation of bamboo could be both sustainable and unsustainable depending on the specie, the nature and form of its operationalization and the political will of government. Effectively, who is championing its goals, where it is actualize and for what purpose such goal is being herald will determine the cultivation. The overarching goal of cultivating bamboo is the eco-efficiency which is an important pathway to an eco-sufficient future. Large scale cultivation will ensure agricultural activities are made not just ecologically sustainable but economically viable. According to INBAR, Bamboos' cultivation remit to promote a post-fossil society in a manner that the generation of waste is not only limited but are used for feedstock in the production of other products.

Bamboos can also be used in the REDD^{+} projects. REDD^{+} is World Bank and UN-REDD programme of providing finance to developing countries that manage or plan to preserve their forests. REDD^{+} also seeks to achieve an overall emissions reduction or stabilization through management of carbon stocks and flows. REDD^{+} goal is to ensure improvements in land management practices across a range of land uses. The purpose is to strengthen the capacity of the ecosystem to sequester and maintain carbon stocks through improvements in a variety of land use practices aligned with local goals and aspirations for sustainable development. An international instrument for sustainable forest management and maintaining biodiversity is the GEF. The GEF is a UN initiative of investing in sectors making a positive impact on the environment and quality of life. It is a pioneering investment to help develop and grow businesses in the South expanding forestry sector by bringing jobs to those communities as well as broader potential ecological benefits. The fund can play a vital role in bringing about a climate resilient civilization if the laborious documentation process is reduced.

Notes

1. Roche, M. Y., Ude, N., & Ofoegbu, I. (2017). True Cost of Electricity: Comparison of Costs of Electricity Generation in Nigeria. A Report of Heinrich Boll Stiftung/The Nigerian Economic Summit Group.
2. NNPC- Nigerian National Petroleum Corporation (2017). Nigeria to earn US $150 m from Bio-fuel Initiative Annually Available at: https://www.nnpcgroup.com/PublicRelations/NNPCinthenews/tabid/92/articleType/ArticleView/articleId/204/Nigeria-To-Earn-US-150m-From-Bio-fuel-Initiative-Annually.aspx. Retrieved on 17th December, 2017.
3. Heinrich Boll Foundation and Friends of the Earth International. (2015) Coal Atlas: Facts and Figures on a Fossil Fuel Available at: www.Nigeriacoal.org Accessed 12th October, 2017.

4. FAO GeoNetwork Database; World Bank 2007–2011.
5. FGN 2017.
6. Roche, M. Y., Ude, N., & Ofoegbu, I. (2017).
7. PVGIS © European Communities, 2001–2012, HelioClim-1 © MINES Paris-Tech, Centre Energetique et Procedes.
8. Cervgni et al., 2013.
9. Schaltegger, S. Burritt, R. and Peterson, H. 2003. *An Introduction to Corporate Environmental Management:* Striving for Sustainability, Greenleaf Publishing, Sheffield.
10. INBAR 2012.
11. Ibid.
12. Kuehl, Y., Henley, G., and Yiping, L. (2011).The Climate Change Challenge and Bamboo: Mitigation and Adaptation, INBAR Working Paper 65, A Publication of INBAR Kuehl, Y. and Yiping, L. (2012). Carbon Offsetting with Bamboo, INBAR Working Paper 71.
13. Ibid.
14. RMRDC 2010.
15. Ogunwusi and Onwualu, (2011).
16. Ibid.
17. INBAR 2012.

Chapter 7
Gender Relations and Forest Resource Management in Post-COVID-19 Age

> Humanity is not a bunch of lemmings marching unstoppably toward a cliff. There is such a thing as free will.... People please wake up! For the sake of young people, future generations, and other life on our planet, don't settle for what some "experts" say is the best we can do.
> – James Hansen

Changing Role of Gender in the Post-COVID-19 Age

The last two decades recorded dramatic changes in the socio-economic milieu of Nigeria. Amongst others factors the need to stabilize forest resources in manner that it sustains present and future generations plays vital role in this transformation. The goal is to maintain the biodiversity, stabilize atmospheric carbon cycle in tackling the menace of climate change while also providing man with a means for sustaining his family.[1] Despite these beneficial attributes of forest, there is a gradual decline of the national forest. With the depletion, forest's role of combating poverty, ensuring food security and providing decent livelihood and delivery of vital long-term ecosystem services such as clean air and water, conservation of biodiversity and mitigation of climate change are threatened.[2] This is because forest management practice has changed substantially. Over the last 25 years, the extent of the world forest continues to decline as human population, urbanization, biomass extraction and demand for food and land have resulted in net forest loss of over 50%.[3]

In recognition of the critical impact of the depletion of the means of sustaining life on the planet, the global community through the Brundtland Commission of 1987 put in place measures to curb unsustainable utilization of forest resources. A key issue uncovered to be instrumental to changing the utilization of forest resources is gender. Gender issues play critical role of either resolving or exacerbating forest depletion. Depending on the gender policy in place in a community, forest could be efficiently utilized or adversely depleted. Since the conceptualization of sustainable development by the Brundtland commission, the global community through the convention on Biological Diversity (CBD) and Agenda 21 has placed on the front

A. S. Okoh, *Oil Mortality in Post-Fossil Fuel Era Nigeria*,
https://doi.org/10.1007/978-3-030-60785-2_7

burner the issue of sustainable forest management (SFM). Since then, other international instruments promote gender equality while also recognizing the gender and climate change link.

Different Charters, Convention and Alliances are targeted at mainstreaming gender perspective into governance of forest resources. Though there has been remarkable improvement in gender relations since the SDGs, gaps in the implementation of these goals still exist. Even though gender aware institutions have proliferated, social differentiation has not been totally eliminated. This may not be unrelated with the different meanings of gender floating around in both academic and political circles. Indeed, the term gender is the most used and least understood or agreed upon term in the field of Forest Resource Management (FRM). Because of this gap in understanding, the phrase gender sensitive has now become a catchphrase synonymous with women's concerns and demands thereby ignoring wider and far reaching ramification of it as it relates to forest resources management. The emerging trend in most countries is to compartmentalize gender into specific role and in the process are using the word to perpetually place women into a marginalized sex needing succour and handout from men in order to be partakers of the commonwealth. Nevertheless, this should not be the case because women are not just keepers of the homestead but are instrumental in maintaining forest resources.

However, remarkable transformation of forest resources management in the post-COVID-19 Age is needed to change how humanity relates with nature. At the heart of this transformation is the interest of different stakeholders who hold different ecological norms and values which is altering humans as we deplete forest resources. A major issue with strong influence on how forest resources are used today is COVID-19. Though different measures have been taken to address this problem, gender issues which is a major determinant of this relationship has not been totally eliminated as gaps still exist in the level of socialization of nature. As a social construct, gender is slanted towards men thus making women to be stereotype within the existing forestry resource management and the means whereby this androcentric agenda is concretized is the use of power that has succeeded in preventing women from equitably participating in the flow of forest resource. Thus, has placed women in silos that require men gifting them participation in resource use. The problem is not so much want of policies formulated to deal with this dichotomous relationship but more of the hegemonic tendency of the externally driven rationalization of nature causing an inconstant role of men in forest management.

To start with, forest is a living system composed of many species of flora, fauna and micro-organisms interacting together in the environment in which they occur.[4] Forest is a type of ecosystem, which is an important component of life support systems. Different species of trees and herbs in the forest provide not only timber, but food, shelter and habitat for wildlife. These living organisms in the forest present man with fuel, food, medicines, herbages etc. Forest also maintain atmospheric carbon balance and is also a biogeochemical cycling agent i.e. Carbon, Oxygen, Hydrogen, Nitrogen, Surphur phosphorus which collectively constitute about 95% of the mass of the living matter on the planet. Forest also provides recreational and aesthetic experience.

With forest importance to man, conflicts over curbing depletion and who should use the forest resources are commonplace as our adverse extractive processes are impeding the regenerative capacity of the forest. The Millennium Ecosystem Assessment released in March 2005 reports that there is substantial and largely irreversible loss in the diversity of life on Earth due to human actions. The Millennium Ecosystem evaluates ability of the ecosystem to provide ecosystem services such as provision of goods (e.g. food, water, fiber and fuel), regulation of natural systems (e.g. climate, water and disease) cultural benefits (e.g. aesthetics spiritual, recreation and education) and other supporting benefits. The Assessment concludes that humans have changed ecosystem in unprecedented ways over the last 50 years. Consequently, over 60% of the world ecosystem is now degraded putting much strain on the natural functions of the Earth such that the ability of the plant's in the ecosystems to sustain future generations can no longer be taken for granted.

Equitable utilization and management of these resources calls for distributive justice with intertemporal dimension. This is in line with the sustainable development. Sustainable development insists on the interrelatedness of social, economic and environment sustainability concepts such that resources are utilized equitably between present and future generation (Fig. 7.1 captures the different components).

Unfortunately, the principle of sustainable development has not been fully embraced as forest resources utilization is skewed toward a gender. Gender relation play crucial role in the management of forest as the power weld by different gender determines not just its conceptualization but also the degree of access to these resources.

It is pertinent to underscore that the term gender refers to the relationship between men and women; the significant differences that exist in most societies between the rights and the opportunities of men and women, including rights to land resources, work opportunities and wages and participation in decision making process. According to a UNDP Report of 2009, gender determines what is expected, permitted and valued in a woman or a man in determined context. It is also the differences in socially constructed roles and opportunities associated with being a man or a woman and the interactions and social relations between men and women. With these different definitions, gender has now become a process of engenderization,

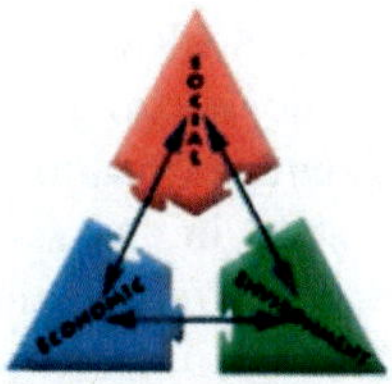

Services	Equity	Biodiversity
Household Needs	Participation	Natural Resources
Industrial Growth	Empowerment	Carrying Capacity
Agricultural Growth	Social Mobility	Ecosystem Integrity
Efficient Use of Labor	Cultural Preservation	Clean Air and Water

Fig. 7.1 Distinct characteristics of the different perspectives to sustainable development

male or female become men or women by acquiring the cultural defined attributes of masculinity and femininity where they take on the appropriate roles and responsibilities for these categories. The understanding of the term is one of it being a constitutive element of social relationships based on perceived differences between the sexes with gender now a primary way of signifying relationships of power. Thus, the usage of gender is both a key force and a process that shapes or 'socially constructs' relationships between men and women. In this way, gender becomes the conceptual "touch" that illuminates the distribution of power within those relationships. And the main avenue for promoting this process of engenderization is the forest. Given this discrepancy, there is the imperative to bridge the gap between genders by placing distributive justice at the centre-stage of all political processes.

The goal of Forest Resource Management (FRM) is to act as a bridge between man and nature based on efficient use at its heart. It insists on inter-temporal dimension with future generation that way forest must be used with moderation to allow its regenerative capacity to sustain it. FRM brings different stake-holders men, women, future generation and nature into a collective harmonious relationship with the stewardship of forest resource as its cardinal goal. The mechanism whereby this harmony is consummated is eco-efficient technologies. FRM requires maintaining the basic structure and function of ecosystems through time for sustainable multiple uses, and to enhancing the quality of the environment to best meet the needs of the society. This entails considering not only the components of an ecosystem, but also the ecosystem processes most especially ensuring the regenerative capacity is sustained.

According to UNDP/UNEP successful management of natural resources depends on freedom of speech, a dynamic and wide-ranging public debate through multiple independent media channels and an active civil society engaged in natural resource issues, because of the nature of the shared resources the individuals who are affected by the rules can participate in setting or changing them.[4] The users have rights to devise their own management institutions and plans under the recognition by the government. The right to resources includes land, water, fisheries and pastoral rights. The users or parties accountable to the users have to actively monitor and ensure the utilization of the resource compliance with the rules and to impose penalty on those peoples who violates the rules. These conflicts are resolved in a quick and low-cost manner by the local institution according to the seriousness and context of the offence.

However, the practice of FRM is quite different from the theory. Discrepancies on whether the economic, social and environment pillars of sustainable development should guide resource utilization and the rules ascribed to different gender are some of the issues at stake in the post-COVID era where all gender feel the same economic burden. Gender differentiation exist which implies there are conflicts. And where there is conflict it means an issue exist needing urgent solution. It then means there is disagreement or something has gone wrong. In the post-COVID Age, local institutions resolve these issues within their locality based on norms and values of the community.

Constraints to Gender Sensitive Forest Resources Management

Forest Resource Management requires an efficient utilization of resources through eco-friendly technology. But this approach can only be successful if gender differentiated responsibilities are reduced, if not totally eradicated. The following are some of the key barriers to gender efficiency in FRM:

I. **Low Participation**—Forest Management is mainly an all men affair with women playing roles that can bring minimal economic benefit. In most Nigerian communities it is generally held that men are more likely to meet the rigors of extraction of resource hence the emphasis placed on them to manage these resources. As a result, there is a low participation of women in forest development, instead women play secondary role are mainly engaged in gathering leaves, herbs, seeds, barks, fruits, nuts, mushroom, root etc.

II. **Commodification of Forest Resources**—Forest resources are increasingly viewed as natural capital capable of generating returns to prospectors. Today, vast economic benefits accrue from depletion of resources with the control of forest vested in men who weld the instruments of state. With the economic fortunes accruing to men, they have succeeded in controlling both women and nature.

III. **Property Right**—Linked to the above issue is the ownership of land. Property right in Nigeria is slanted towards men. Through the patriarchal nature of our society male have succeeded in dominating the ownership of land. The right to forest in most communities is from inheritance which is targeted mainly at the dominant gender.

IV. **Lack of Essential Knowledge**—Education plays critical role in Forest resource management. Women productivity in forest is constrained by the lack of access to essential knowledge in forestry most especially their benefits in practicing forestry. Women and men have different knowledge about forest related issues which also informs why society has limited their roles to different areas of their expertise.

V. **Cultural/Religious Beliefs**—The Forest provides house-holds with essential requirements, religious beliefs and cultural norms are the major determinants of its utilization, conversation and preservation. Some of the religious beliefs and oppressive cultural norms and values discriminate against women in the practice of forestry.

VI. **Androcentric Environment**—The environment by extension forest is seemed from the male perspective. Different institutions and policies in Nigeria have structured women into playing traditional reconfigured roles to suit a gender.

VII. **Impact of Globalization**—Globalization, trade liberalization and some economic policies support export production which are rapidly drawing on the Forest to sustain the trade. Consequently, most Nigerian societies are engaged in the global trade through production of cash crops such as palm oil, rubber and cocoa. This marketization drive has succeeded in drawing these communities into the orbit of capitalist accumulation and in the process is not only creating inequalities but is causing social difference including gender and class.

Factors linked with the failure of resource conservation are shown in Fig. 7.2.

VIII. **Institution and Coordination**—Institutional weakness exist in most institutions vested with the responsibilities of FRM. And this is the root cause of gender bias in resources management which favour depletion of forest resources. The different regulatory agencies in Nigeria such as the Federal Ministry of Agriculture, Department of Forestry, Ministry of Women Affairs have differing approach towards resolving male/female in balance. With the lack of coordination between the different arms of government, marginalization is rampart. In-deed, there is also lack of coordination between NGOS, CSOs and forestry department in FRM.

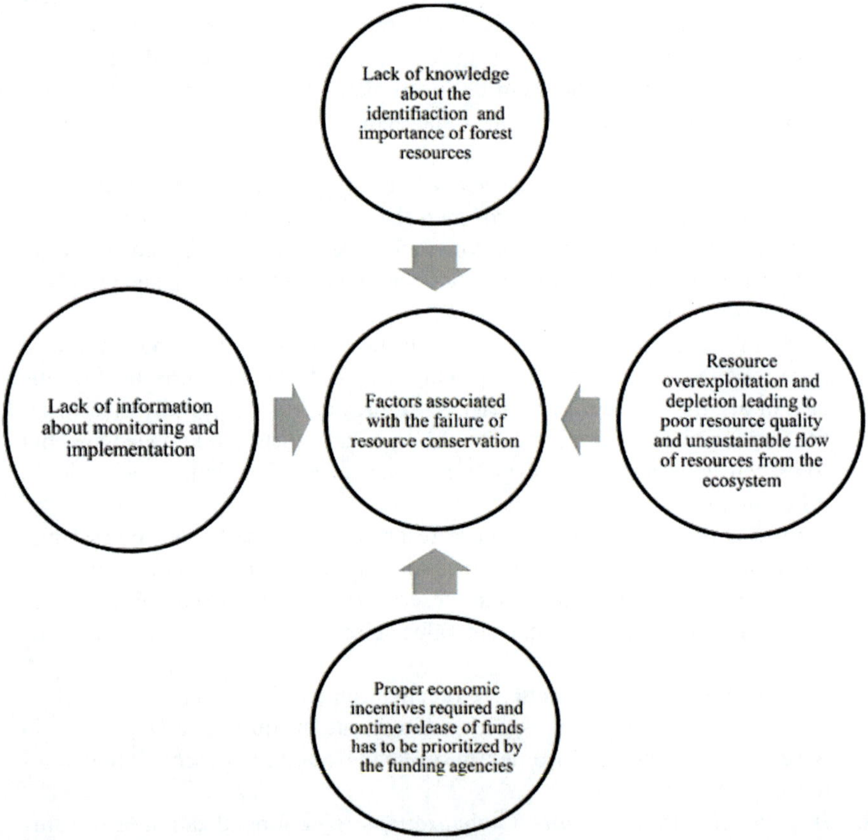

Fig. 7.2 Factors linked with the failure of resource conservation. *Source* https://www.earth-policy.org/indicators/C56/forests_2012

IX **Participation in Plan**—Another grey area in FRM is the lack of involvement of women in the planning stage of issues such as tree planting or decarbonization. This is widening the gap between women and men to the extent that gender issues are considered only as reconciliatory approach not borne out of necessity.

Gender and Forest Resources Management in the Coronavirus Age

In the last two decades, Nigeria's forests exploitation is accelerating dramatically. Though different approach has been adopted to curb depletion and biomass consumption, the rate of degeneration is still growing. Exponential population growth expected to be over 400 m by 2050 will place additional burden on the depleted forest. The institutionalization of gender provides an avenue for examining the interrelatedness of theory, policy and practice within the area of environment, and forest resources management, where gender is a politically negotiated and contested issue involving social relationships. Correspondently, mainstreaming of gender into the national development policy is catching on with gender sensitive issues addressed along lines laid out in the SDGs across most sectors. Nevertheless, gender sensitive development as it relates to forest resources management has lost its critical edge due to androcentric influence. Thus, it has made its effectiveness as a transformational tool to be reduced. As a result, gender is now equated with the study of women and their roles. But gender transcend such narrow compartmentalization since women play a central role in most facets of life as they are efficient managers of resources and therefore should be key actors in natural resources management programmes.

More-so, that men and women hold differentiated interests in FRM in the post COVID world through their different roles, responsibilities and knowledge. Instead of their diverse understandings playing vital roles in sustainable forest management, the dynamic of gender is used to promote male dominated power relations. In most rural communities, women are better suited as manager of resources and are usually assigned reproductive roles which explain why they are the main collectors of forest products and food for their families. This critical role of maintaining the biodiversity must be sustained in the post-COVID-19 age if we are to avoid depletion of endangered species capable of causing another pandemic.

Unfortunately, development planners interpret this to mean that women should be targeted in conservation projects since their daily chores connect them more closely to natural resources. In some cases, gender is primarily relevant only within households and is non-existent within political circles and decision-making processes thereby undermining their inputs and roles in FRM. Attempts at mainstreaming gender into development activities such as the economic stimulus package instead of curbing the dominance of men has subordinated women to secondary role who are gifted some concessions to create the ambiance of a level playing ground between all sexes. By giving social concerns of women new attention in the restoration of forest it has resulted in women's participation which only draws on a view that women are principal fixers of the degraded environment. Whilst in the real sense, it is nothing but a ruse to deliberately mobilize women into rendering unpaid labour to meet the demands of conservation projects.

For this reason, gender is a social construct which serves to perpetually marginalize women in the distribution of forest resources and are only called upon when there is adverse degradation. So, the attempt at inserting gender issue into development plan through mainstreaming into some facets of economic stimulus like the feeding of school children during the lockdown is not properly conceived. What this has achieved is that it has placed gender disparity at the fore-ground of all discussions. Thus, we have theoretically inserting gender concerns across policies and development practices at different levels without addressing the root cause of the problem.

Hence, we can say that gender mainstreaming has never been even. What has emerged is that the unevenness reflects the differing levels of commitment and resources invested across gender. It is also a means of ensuring that issue of gender is depoliticized and made technical. Effectively, women groups are incorporated into national policy framework to realize conservation goals without addressing the complex politics of gender-equality. Unfortunately, Nigeria is a patriarchal society where the male gender dominates decisions making processes. Consequently, decisions on forest resources are tailored towards their needs with men's priorities determinant of resource management. For instance, small-scale exploitation of non-timber forest products (NTFP) used for handicraft or for invigorating purposes are predominantly conducted by males whereas, the gathering of firewood and consumables NTFPS are tasks assigned to women.

Stereotyping plays another crucial role in gender issue in the post-COVID-19 world. Due to stereotypes, men are more likely than women to get trained in the latest technology of managing forest. This training succeeded in positioning men to get the best jobs in FRM that uses advance technologies. Again, there is difference between men and women relationship to the means of production. Depending on the existing social relations of production, male have more access to the means of production thus ensuring their unending hold on power. Patriarchal character of society helps to institutionalize inequality and this also has affected the farming systems which relegated women to some specific activities.

The Paris Agreement brought the need to stabilize atmospheric temperature to below 1.5 °c threshold. A medium for curbing depletion of forest resources is $REDD^+$. This UN sponsored $REDD^+$ program aims at ensuring that rich countries provide finance to developing countries involved in preserving their forests. $REDD^+$ seeks to radically achieve the overall emissions reduction or stabilisation through the management of carbon stocks. Ultimately, then, $REDD^+$ overall objective is emissions reduction through the management of carbon stocks and flows and by doing so improve land management practices across a range of land uses. Hence, the scheme's goal is to enhance host communities' capacity to sequester and maintain carbons stocks through improvements in a variety of land use practices aligned with local goals and aspirations for sustainable development.

But the introduction of $REDD^+$ mechanism reflect male interest in preservation which is contrary to the conservation which is the reality on the ground in most communities. The way and manner $REDD^+$ is implemented only serves to compensate those primarily responsible for the problem of climate change. What $REDD^+$ has succeeded in achieving is to criminalize the host communities for carrying out activities that support their families. For instance, in Cross-River State the $REDD^+$ mechanism is criminalizing women for gathering wood and carrying out their traditional forest-related economic activities whereas some male dominated interest still carries out logging activities in forest designated for $REDD^+$.

In its Nationally Determined Contribution (NDC), Nigeria agreed to address anthropogenic climate change through mitigation action of stopping charcoal production. Effectively, women who are involved in this trade will be in for hard-times. One would expect that government should have put in place an alternative source of energy before proceeding on the remarkable steps which will have significant impact on the environment. But without any alternative, the goal of meeting this commitment may be political rhetoric as fuelwood will continue to be extracted.

The Post-COVID-19 FRM We Want

Gender issues will play an important role in FRM in the post-COVID era. When we talk about gender issues here, we are talking about the way different sex have a say in forest governance and how different sex contribute to sustainable resource management such that there is equity in resource use between different gender and with present and future generation. Nigeria's approach to addressing the issue of gender imbalance is promising. Today, women are more visible across most sectors and are now playing roles traditionally the responsibility of men. Nevertheless, women's role in the MFR has not really changed. There is stereotyping of forest management as men are still the dominant sex. The present forest regime is a short-term coping mechanism which does not build resilience of the vulnerable class most especially women and children who are more prone to the challenges of climate change and COVID-19. For inclusive green growth, women and children's priorities should form the bed-rock of any post-Paris or economic recovery solution. This is because low carbon emissions development or economic recovery will be more effective and more equitable when they are designed using a gender-informed approach.

More so as millions of women in the country make decisions daily that influence the amount of carbon released into the atmosphere. As home-makers, women's choice can be changed such that it influences the regeneration of forest thus sustaining both biotic and abiotic components of the ecosystem. This requires a cultural shift which reduces their carbon footprint, through gender sensitive approaches that redistributes resources to stem their adverse interference with bio-geophysical properties of the forest. Furthermore, strengthening women's political representation and increasing their leadership roles in society will ensure that women priorities are not only implemented but will improve their economic circumstance. There is the imperative for institutional transformation aligned towards poverty alleviation with social safety nets created to empower women. Schemes such as Anchor Borrower Scheme, N-Power and other palliatives should be encouraged. Women and female child's ability to participate in education, productive and civic activities should be intensified. The time spent by women and girls collecting fuelwood and water collection can be reduced drastically by the provision of accessible and affordable transport systems, energy and water and sanitation systems.

There is need for capacity building to increase women's chances of finding employment with training to boost sales of goods outside the village, thereby increasing income and expanding their social networks. Improving women's access to alternative sources of energy other than traditional bio-fuel can reduce women's time burdens, their exposure to indoor air pollution, and the risks of other health problems. Cooking fuels such as kerosene and LPG are good substitutes but where this is not feasible other interventions such as clean energy cooking stoves be distributed.

An efficient FRM should be all encompassing to cover all relevant sectors in order to maintain the biodiversity (This is shown in Fig. 7.3) to ensure an inclusive green growth.

From Fig. 7.3 it is evident that resource management affects the overall quality of life including human health. Women's active participation in biodiversity conservation programs is essential for resource management to be effective and cost-efficient. Promotion of Good governance is bedrock of efficient forest resource management.

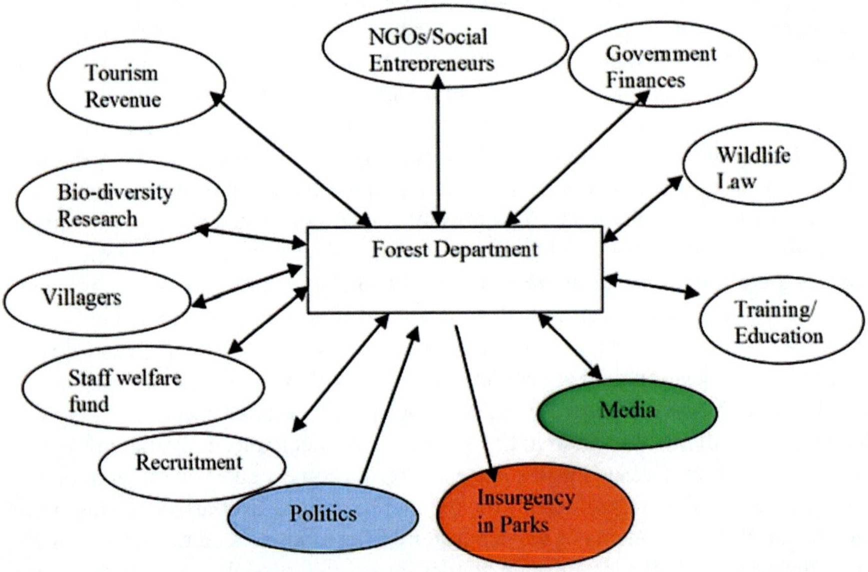

Fig. 7.3 An all encompassing FRM covering all sectors. *Source* www.iimahd.ernet.in/egov/ifip/susan-sharma.htm

Good governance is a vital component of sustainable development as it is the mainstay of all socioeconomic and political interactions. An inefficient government will result in a failed state while an efficient government creates enabling environment for sustainable development.

To breathe a new life into FRM, an approach I call the 'Food Sufficiency Economy' (FSE) is suggested. The FSE concept fills the gap in utilization of fuelwood in a biomass economy. The concept's appeal lies in placing a barrier on corporate malfeasance and profligate squandering of resources. In this theorization, prejudice common to many societies is resolve through nurturing a socially inclusive growth. The goal is to plot a path towards imbibing environmentally sound principles like Africa's eco-biocommunitarianists' world view. Within this blueprint, there is harmonious relationship between human and nature since all species have co-equal value. This also includes women since it is based on people right to determine their energy and water right. It is in line circular economy but a home-grown solution based on Africa's cultural values. In the blueprint, agricultural transformation based on Climate Smart Agriculture (CSA) will give birth to a new culture of natural resource stewardship which will gradually replace the current unsustainable production in the post-COVID-19 Age. In the post-COVID-19 age countries will be self-sufficient hence not reliant on others for their food and fibre.

Notes

1. Okoh, A.I. S. (2013). Political economy of climate change in Africa, Lambert Academic Publishing. German. Available at: https://www.amazon.com/Political-Economy-Climate-change-in-Africa/dp/3659373117.
2. IEA (International Energy Agency) (2018). World Energy Outlook 2018 examines future patterns of global energy system at a time of increasing uncertainties [Online] https://www.iea.org/ Retrieved on 1st February, 2019.
3. IEA (2014). Africa Energy Outlook: A Focus of Energy prospects in Sub-Saharan Africa. A Publication of International Energy Agency Paris, France. Available at: https://www.iea.org.
4. *UNDP, UNEP(2005), The World Bank and World Resources Institute – The Wealth of the Poor: Managing Ecosystems to Fight Poverty Institute 2005 Chapter 3 The board's role in governance, sources 2005".*

Chapter 8
Green Economic Recovery in Post-COVID-19 Era

> By combining measures tailored to specific national circumstances, governments can both support economic recovery from today's crisis and underpin the energy transition required to avoid a climate crisis in the future.
> Energy Transition Commission, 2020

Economic Stimulus for Post-COVID Era

In April 2020, the global energy market recorded unimaginable shock with benchmark futures trading for less than zero in the US while Brent oil traded at an all-time low of $15. Although Nigeria's Bonny light did not fall to the negative threshold but there are massive losses from every barrel the country produced. Implication for Nigerian oil dependent economy is immense. With the devastations of coronavirus disease, Nigeria's monocultural economy slipped towards the worst economic collapse that include a devalued currency and soaring inflation as oil prices go negative.

COVID-19 is the defining challenge of modern age. It is not only a public health emergency but is equally an economic crisis. It is bringing personal tragedy into households across the globe, and an economic crisis. GDP and government revenues are falling as unemployment and poverty rise. Stimulus packages target sectors deflated by coronavirus externalities. They are typically designed to mobilize money quickly; ensure that expenditure is temporary; and target a specific problem such as recapitalizing banks or boosting consumer spending. Accidentally, stimulus is placing funds where is ought but in some cases dumped money into the real economy guided by panic rather than purpose. In COVID-19 time, a unified moral purpose should guide global efforts of building a low-carbon, climate resilient and inclusive economy. To this end, the world's largest economies have begun to design and deploy economic stimulus measures that will surpass €5 trillion. This includes a €1 trillion package proposed by the European Commission. In Nigeria government propose a stimulus package of 3.5 trillion Naira. This is just the beginning. Additional expenditure will come as the crisis deepens and how we spend this money will shape our world for at least a generation.

A. S. Okoh, *Oil Mortality in Post-Fossil Fuel Era Nigeria*,
https://doi.org/10.1007/978-3-030-60785-2_8

COVID-19 forcefully placed on the front burner inadequacies of the world by putting public health issues at forefront of national budgets. It also brings the failure to build a resilient economy and society into focus. In the drive toward rapid industrial growth, governments underestimated re-emergence of SARs epidemic hence under-invested in healthcare system. Neglecting to address intersecting political, social and economic inequalities exacerbated the pandemic. These are challenges the stimulus packages must address if societal resilience is to be achieved. This means readjusting priorities with renewed focus on health and unemployment benefits, labour standards, and social care services.

The stimulus packages also must tackle constraints felt small holder farmers resulting from the coronavirus. Not only are small holder farmers the most vulnerable class but disproportionately bear burden of the pandemic. Most family farmers are not just poor but are within the population of those living on less than $1 in 2020 as evident from the National Bureau of Statistics findings of 2020. Poverty is a not a just a threat multiplier but reinforces risks creating further hardship for the people.

China was in the forefront of providing support to many African countries and making donation of items as well to many African countries including Nigeria. The European Union and some of the member states were also supportive and shown solidarity with African countries in the wake of the COVID-19 pandemic. Germany for example pledged to support Nigeria with 5.5 million Euro to help the country in the fight against Coronavirus pandemic while European Union donated Euro50million for the same purpose. India smoothen further their relationship with the United States when they supplied pharmaceuticals to the United State at the request of Donald Trump to tackle the Coronavirus. In the words of Donald Trump, he called Indian President a trusted ally and a friend indeed because India supported the United States in the time of need.

Philanthropists across the world, the private sector, banks, civil society organizations, Non-governmental organizations, faith-based organizations, research institutions etc. stepped up their support at international, regional, national and local levels in the fight against COVID-19 pandemic. The supports came in different forms, shapes and sizes. In some cases, these group form alliances, groups and the likes to deliver their support. In many African countries, the Chinese philanthropist Jack Ma was noted to have made donation of equipment especially the personal protective equipment (PPE) across the continent. Beside the support made to other African countries, Aliko Dangote also chaired the private sector led group of the Coalition Against COVID-19 (CACOVID) in Nigeria. The coalition raised several billions of Naira from all walks of life in support of the Federal Government of Nigeria's fight against COVID-19 pandemic.

In May 2020, the Central Bank of Nigeria strengthened the Nigerian economy by providing a combined stimulus package of about N3.5 trillion in targeted measures to households, businesses, manufacturers and healthcare providers. These measures are deliberately designed to both support the Federal Government's immediate fight against COVID-19, but also to build a more self-reliant Nigerian economy. To bring to fruition government's economic stimulus package the CBN highlighted the steps Nigeria must adopt as follows:

(1) Build a base of high quality infrastructure, including reliable power that can engender industrial activity;
(2) Support both smallholder and large scale agriculture production in select staple and cash crops;
(3) Create an ecosystem of factories, storages, and logistics companies that move raw materials for value-added production, and finished goods to markets;
(4) Use our fiscal priorities to create a robust educational system that enables critical thinking and creativity, which would better prepare our children for the world of tomorrow;
(5) Develop a healthcare system that is trusted to keep all Nigerians healthy, irrespective of social class;
(6) Facilitate access to cheap and long-term credit for Small and Medium-Scale Enterprises (SMEs) and large corporates;
(7) Develop and strengthen pro-poor policies that bring financial services and security to the poor and the vulnerable; and
(8) Expedite the development of venture capitalists for nurturing new ideas and engendering Nigerian businesses to compete globally.

Green Economic Recovery Tools in Post-COVID-19 Era

The year 2020 was to mark a turning point in global effort to reduce emission of GHG with different activities line up for the big year of climate action. The year marked five-year point at which countries were to step up their climate ambitions, with expectations running high that COP26 (which was scheduled for November, 2020 in Glasgow) could have outcomes as consequential as COP21's Paris Agreement in 2015. But all this changed with Corona virus. Uncertainties associated responding to the immediate health and financial crises posed coronavirus pandemic made the event be shift to 2021. Indeed, uncertainties pervade COVID-19 making it difficult for one to state with a high degree of confidence outcome of the pandemic. One thing, however, is certain: Globalization in its past reincarnation is dead. In its place is a new world order emerged. The world order is slant towards isolationists values amplifying need for countries to seek local remedies for all forms of requirements. This new world order will be one where countries look inward for food, health and education. COVID-19 avails the country opportunity to travel on a new path of recovery, one targeted at making the country great again. The new country emerging from the trauma of COVID-19 is one where economic reasoning of nature is modified to create equitable socialization of human and non-human resources. The post-COVID-19 age is a time for transforming Nigeria into a modern, sophisticated and inclusive economy. This is a society that is self-sufficient, protects the poor and vulnerable, and can compete internationally across a range of strategic sectors.

Yet, it is not Uhuru for the global economy neither has Nigeria come out of the woods of coronavirus pandemic. Though COVID-19 is peaking in some economies others are experiencing further devastations of the pandemic. The result is various countries moving on diverse trajectories on mitigation the virus. But the consensus amongst countries is for the easing of the lockdowns to allow economic activities needed for generation of revenue to rise thereby upturning the economic downturn. As health concerns lessen, however, countries are now focusing on stimulus packages to stem the economic impact of COVID-19. As they do, it will be vitally important to help them build the foundation for a more resilient, sustainable, and prosperous future. This foundation can only be effective if it is aligned with current wave of technological shift gaining traction lately.

This is because two inversely related issues drive global urgency to reprioritize economic activities: The Coronavirus pandemic and fall in oil price. Though these issues are mutually exclusive they are also reinforcing one another. For instance, coronavirus is driving down the price of oil with most countries either in partial or full lockdowns. These threats will stifle economic activities in the process impairing realization of the objectives of governments. With adverse health impact of COVID-19, major economies are drawing up enormous economic packages to cushion shocks of coronavirus pandemic. This palliative is meant to ameliorate effects and promote a healthy environment for the economy to be back on track.

The effort presents unique opportunity to governments' drive for shift to a low-carbon future while also growing the economy to meet other emerging crises in the future. One the tools flaunt by world leaders to meet these goals and ensure sustainable human health is green economic recovery. Green recovery is adjudged solution to the perplexing issues of reviving the economy while also addressing climate change. The approach was suggested by German Chancellor Angela Merkel and International Monetary Fund Managing Director Kristalina Georgieva in April 2020. Since then, the concept has emerged as a political fault line from the United States to India and South Korea.

A study with some of the co-authors as the Nobel laureate Joseph Stiglitz and Lord Nicholas Stern examined the stimulus for "green recoveries" from COVID-19 fleshed out around the world. It found that the pandemic presents the world a new opportunity to reorder priorities stating: "the COVID-19 crisis could mark a turning point in progress on climate change," the authors wrote. Adding, much of the change in carbon consumption would depend on policy choices made in the next six months. Hence the imperative for green projects that boost renewable energy or energy efficiency which will create more jobs, deliver higher short-term returns and lead to increased long-term cost savings relative to traditional stimulus measures. But according to Matthew Green in the Reuters of 5th May, 2020,[2] there were some risks with extrapolating from past crises to discern how coronavirus recovery packages might play out, given the possible reluctance of people to travel or socialize following the pandemic.

Different governments all over the world have focused on emergency economic relief as an estimated 81% of the world's workforce has been hit by full or partial lock-downs. But as governments move from "rescue" to "recovery" mode, the clean energy sector could provide strong returns in terms of both rebooting economies, creating jobs and advancing climate goals. This informed a shift to cleaner energy sources. In industrialized countries it was the suggestion that they should focus on backing "clean physical infrastructure," such as solar or wind farms, upgrading electric grids or boosting the use of hydrogen. The study also recommended retrofits to improve building efficiency, education and training, projects to restore or preserve ecosystems, and research into clean technologies. Whilst in lower- and middle-income countries, support for farmers to invest in climate-friendly agriculture came out ahead in the study. With carbon emissions on track for their biggest fall on record this year, governments have the opportunity of choosing to either pursue net zero emissions targets or lock in a fossil fuel system that would be "nearly impossible to escape."

Economic Stimulus Package in COVID-19 Era

Economic stimulus packages established by governments around the world will be vital to sustain economic activities, jobs and livelihoods over the next few years. The current crisis reduced the already low interest rates at which advanced economy governments can borrow, while it increased the risk premium paid by many private sector players as well as countries perceived as riskier. In that context, governments are in a position to drive a massive wave of investment to stimulate the global economy, either by using their own balance sheets or reducing the cost of private capital through various forms of financial support. Indeed, trillions have already been committed to support immediate economic stabilization and longer-term economic recovery. Development finance institutions will need to play a crucial role in ensuring such action extends to developing countries facing higher risk premiums.

The lessons learnt from the COVID-19 crisis is it dramatically exposed unpreparedness of the global economy to systemic risks, despite early warning from scientists. In 2019, climate change was linked to at least 15 extreme weather events costing between US$1–10 billion each.[3] The IPCC predicts that such extreme weather events will likely become more frequent with the rise in global temperatures.[4] Investing in high-carbon activities without climate conditionality in the hope it will help the global economic recovery would only prepare the ground for future systemic crises. Economic stimulus packages should contribute to building a healthier, more resilient, net-zero-emissions economy.

According to the Energy Transitions Commission (ETC) (2020), companies and organizations hit by the economic downturn should continue to be committed to building a better economy. This should be one enhancing the health of people and ecosystems, and that builds our resilience by better anticipating and mitigating risks. In particular, it is vital to reinforce action to address the climate change related risks, by achieving net-zero greenhouse gases emissions by mid-century and driving sustainable economic prosperity. That, clean energy, low-carbon and digital solutions are fundamental pillars of a better economy: they can improve the quality of the air we breathe, enhance our quality of life, and limit the occurrence of climate-related disasters. They can also underpin new businesses and new jobs: according to IRENA, the cumulative gains for transforming the energy system could reach US$98 trillion between 2020 and 2050, greatly exceeding the related investment costs US$ 15 trillion.[5]

Governments have the choice, the power and the responsibility to build this new economy faster with businesses and civil society. The ETC believe they can do this and help the global economy recover by putting 7 key priorities at the heart of economic stimulus packages.

These are[6]:

1. Unleash massive investment in renewable power systems
2. Boost the construction sector via green buildings and green infrastructure
3. Support the automotive sector while pursuing clean air
4. Make the second wave of government support to businesses conditional to climate commitments
5. Provide targeted support to innovative low-carbon activities
6. Accelerate the transition of the fossil fuels industry
7. Don't let carbon pricing and regulations spiral down.

Clean Energy Transition

Global carbon-based growth together with the novel coronavirus has brought clean development to the foreground of political and economic discussions. Some few years back, attention was on issue of decarbonization of the global economy which gravitates around sustaining green infrastructure to meeting economic growth requirements. But as of late, this growth-based mentality is losing traction. The major factors propelling world leaders towards embarking on net zero emissions is hinge on health challenges tied economic growth and agroecological depletions which are now exacerbating human and climate risks. The rising spate of COVID-19 alongside climate change warrant reordering our lop-sided ecological integrity. Despite global commitment, human influence on the climate system is clear, and recent anthropogenic emissions of Greenhouse Gases (GHG) are the highest in history as increased use of fossil fuels have reversed the long-standing trend of gradual decarbonization. Warming of the climate system is unequivocal as GHG emissions are still far from stabilizing requiring deep decarbonization pathways.[7]

Emissions from burning of fossil fuels with direct carbon dioxide (CO_2) emissions from industrial processes has continued to grow from the first Conference of Parties (COP1) of the United Nations Framework Convention on Climate Change (UNFCCC) in 1992 to the current COP26 postpone to 2021. In line with this, reorganization of consumption and production are underway to correct uptake of GHG. However, these circumstances are shape by decisions about production systems and investment priorities in the post-COVID-19 era. This is intermeshed with political manoeuvrings in an increasingly artificial, crowded and changing biosphere.

In this age of the coronavirus, decoupling energy use from gross domestic product (GDP) and population growth is critical in the years to come if global commitment of maintaining emissions at 2 °C threshold must be realize as the current rate of decarbonization needs to double to achieve the global climate goal. Again, most economies are enmeshed in the economic quick sand of growing their economies to withstand the vagaries of climate change and human health security impediments. Given this, clean energy is fast catching on with various organizations and states exploring avenues for embarking on low carbon trajectory. Rebalancing global energy production such that carbon-based growth is scale down is the ultimate stabilization goal of clean energy transition.

A critical component of this stabilization and abatement mechanism deemed capable of rebalancing energy priorities is the 'clean' or 'green' energy initiative. Clean energy pathways can ensure efficient allocation of resources while fostering green growth but can be hijack if not properly managed. At its basic level, clean energy solution is a carefully flesh out low carbon emission trajectories countries can travel towards low carbon emissions. It presupposes eco-efficient utilization of green infrastructure will usher in a dawn of eco-equilibria relations wherein greenhouse gases emissions are limited if not outrightly de-emphasized. It attempts to answer all the needs of man including energy security, food security and climate change: These are three of the major global security challenges humankind is witnessing. With vast clean energy potentials, exploring the abundant energy sources is critical for Nigeria's economic growth and green recovery in post-COVID age.

So, the vehicle for transitioning to an ecologically stable energy future in the post-COVID era is eco-efficient energy supplies adjudged solution by world leaders like Germany's Angela Markus and Britain's Boris Johnson for tackling COVID-19 pandemic at the same time with climate change and agro-ecological contradictions. The instrument wherein this is sustainable is clean energy technologies. But such transition for Nigeria is an endeavour of many contradictory parts. Not just that diverse interests have differing motives for the environment but Nigeria also has serious motives for depletion of her stock of green infrastructure. This is because in Nigeria like the rest of sub-Saharan Africa (SSA), nature is a large bowl for the take-off of economic growth. It is also a concrete pathway to economic and political sovereignty. However, such economic security based on neo-extractive mentality is externalizing costs to others in so doing reinforcing climate-related extreme events. This also informs why current attempt at limiting pollution to net zero emissions is a highly complex endeavour.

Constraints to Clean Energy Future in Nigeria

The barriers to transitioning to a clean energy future are:

Energy Poverty—A major reason for Nigeria's complex energy transition is the mere fact that the country's population is over 180 million with over 120 million living without electricity. Furthermore, Nigeria is sparely populated hence building electricity grids across such distances to reach families living on \$4-\$5 a day does not make economic sense. There is extreme poverty with over 45% living below the poverty line of \$1.2 a day.

Biomass Economy—In Nigeria, over 60% of the population are reliant on biomass for domestic purposes. In this country, biomass is the primary energy supply and of this, solid biofuel accounts for 81.2%. Hence, the energy sector vital in green growth is still evolving. This sector is equally critical to all decarbonization of the ecospheric processes. Making a predominantly rural population attain eco-friendly technologies will have disruptive implications for their livelihoods. Hence biomass economy is not an alternative source to clean energy civilization but is Nigeria's clean future. Nonetheless, biomass also has problematic ramifications for the future if exploited the way it is now. Given that the contemporary biomass economy exacerbates environmental degradation, it is also fuelling hunger. This is because the current biomass-based fuels cause deforestation, flooding, inadequate replacement of tree cover, soil erosion, conversion of cropland to biofuels in addition to other problems.[8] Finding an appropriate long term regulatory measure and energy mix which favourably explores Nigeria's vast biomass potential will require new policies emphasizing energy efficiency and clean technologies while ensuring food security.

Price of Renewables—Nigeria's vast renewable energy potential is largely untapped not just because of the difficult terrain but the cost of implementing clean energy technology is high. Though prices of renewables are falling, it is above what Nigerians can access. The result is there are vast untapped clean energy sources. A clear evidence of this untapped potential due to its expensive nature is in hydropower. Hydropower is immense. About 60% of the population lack access to electricity, while about 70% lack access to clean energy for cooking. Large-scale hydropower has its own problems in terms of inundating sensitive areas, population displacement and siltation. Other sources of renewable energy such as wind and solar are unexplored all over the country. Even where extraction is minimal, the technology is still expensive. Transitioning to clean energy future is a complex endeavour capable of fuelling unsustainable practices that will result in water, energy and food insecurity.

Food and Biofuel Complex—There is the paradox of promoting food crops which are also feedstock in bioenergy production. Fears have been expressed (and accurately so) of the potential effects substitution of food crops for feedstock in land allocation can create. But this does not hold waters where land is not a constraint to increased production. Large scale production of biofuels is likely to cause an increase in input costs for livestock production. The effect comes from the likely increased conversion of pastureland to cropland with the incentive of higher returns from land put to biofuels. Biofuel mandates from EU and US projections for 2015 will cause an increase in global crop cover by 7.09 million hectares of which 3.06 million is from sub-Saharan Africa alone.[8]

Given this, the global prices of food stuffs are likely to increase and alter trade patterns for coarse grains, oil seeds, crude and refined vegetable oils and livestock products. This is against the backdrop that the US alone is the major exporter of coarse grains and is expected to reduce net export of coarse grains by $495.1 million and increase the net export of oilseeds by $960.6.[9] In this era of climate risks, these shifts are likely to have an impact on African economies.

Technology—Technology is the most vital mechanism in a clean economy. It is the medium for transitioning to an ecologically sustainable zero carbon future. But net zero emission is a paradox with both negative and positive implications. The challenge for Nigeria is not so much how to meet her energy requirements but how to achieve such goals despite clean technologies. Clean energy requires high skill which is of limited supply in Nigeria. This state-of-the-art technology underscores its price. Acquiring new skills in a global economy dominated by powerful energy corporations is the defining crisis of energy security. How to balance interests of economically powerful lobbies in a global economy ravaged by climate risks has proven to be man's Achilles heel.

Fuels Climate Change—Ameliorating harmful impacts of climate change is a major driver of clean energy development. The instruments which instill energy sufficiency carry with them the seeds of agro-ecological deteriorations. For instance, Kyoto Protocol instruments such as the Clean Development Mechanism (CDM) potentially can reduce climate hazards. This also includes Reduced Emissions from Deforestation and forest Degradation (REDD) programmes and carbon offsets schemes wherein an organization can buy or reforest a standing forest as offset for their emissions. However, there is conflict of reconciling global emissions with national emission budgets for each state at varied levels of carbon emissions and industrialization. With this impasse, harmful pollution of the biosphere is growing daily creating severe economic and political challenges.

Still, post Kyoto negotiations have not really helped matters rather has exposed the lack of political will by dominant economies to end planetary peril. To Nigeria, the Lima (COP20) resolution to cut emissions based on Intended Nationally Defined Contributions (INDC) is not highly desirable. The country will be more amenable to a carbon cuts model not detrimental to the growth trajectory. But all this is now changed with COP21.

Water and Food Insecurity—Finding an appropriate long term regulatory measure and energy mix which favourably explores Nigeria's vast clean energy potential will require new policies emphasizing efficiency in water and food production. This is because there is a severe agro-fuel complex with food security and food sovereignty imperatives. Food/fuel crops contradictions threaten clean energy development as bio-energy competes for food crops as feedstock.

Land Use Changes—Large scale Land acquisitions are pushing smallholder farmers to marginal lands posing severe stress to the land. Land use changes resulting from over exploitation of farms imply that environmental hazards will contribute significantly to forest degradation as they drive land owners away from their traditional lands into new, forested areas where the only income to be earned is from slash-and-burn agriculture and hunting for bush meat. Increases in urbanization and industrialization have also raised the demand for wood products, especially firewood and charcoal. Consumption of forest products has doubled in Nigeria in the last decade. Unless energy alternatives to firewood is found, as well as alternative sources of income for people whose lives depend on forest consumption, deforestation will continue to accelerate, enhancing climate risks.

Policy Inconsistency—Governments can potentially play crucial role in resolving energy security impediments by redirecting policy and by formulating institutional frameworks for deployment of clean energy. Political commitment and support of government gives policy direction for development of necessary regulatory instruments for advancement of clean energy. Conducive policy instruments engendering development of clean technology are an important component of clean development. Hence all stakeholders must be carried along to attain socially inclusive green growth. But for now, policy inconsistency impairs attainment of clean energy objectives of governments.

Investors' Apathy—Foreign Direct Investment (FDI) is the key to a carbon free future. Investor's apathy may stem from the ease of doing business and exorbitant cost of take-off of business. Coupled with this is the issue of power generation. Again, this adds severe burden to new clean sources. For instance, Nigeria has the laudable goal of reforesting the frontline states in the Northern fringes of the country. To this end, the Great Green Wall Sahel and Sahara Initiative (GGWSSI) initiative aims at limiting desert encroachment through reforestation activities. As laudable as the objective of the project may sound, a shortfall in FDI stalled full implementation is a serious barrier to attainment of its goals.

Governance Deficit—Central to the issue of clean energy transition policy shortfall is not so much the outcome of market failures foist by inappropriate market mechanisms to address environmental challenges but the challenge is in the nature and perception of the state in Nigeria. Indeed, the state is an odd mix of diverse contending forces. Rentier economy resulting from insatiable rent seeking preoccupation of the political class is the driving force for the failure to institutionalize appropriate building blocks for green society. A good example of clean energy vision derailed by neo-patrimonial character of the state is award of clean energy stoves.

In October 2014, the Federal Executive Council awarded a contract for N9.2 billion worth of cooking stoves and wonder bags for rural women under the national clean cooking scheme. Then, it was agreed that 750,000 units of clean stoves and 18,000 units of wonder bags were to be supplied by the designer from South Africa. The scheme was expected to have a lifespan of 5 years for the provision of 20 million stoves. Environmentally speaking, this is sound reasoning as biomass dependency would have drastically dwindled. In addition, this policy has co-benefits of not only creating jobs when assembled locally but would have environmental benefits in the form of carbon offsets thereby earning carbon credits for the country. But with neo-patrimonial politics in place, the sharing of national income is a zero-sum game where the political class is bent on accumulation of both wealth and power. Consequently, the project has not taken off even when the funds were released.[10] What has emerged is graft stalled distribution of clean stoves. In this setting, it will be very misleading to attribute failure of providing clean stoves for rural women on market imperfections alone rather governance deficit is the bane and the problem.

Intermittency—Concerns have been raised over the issue of intermittency of renewable energy. Serious doubts expressed on regularity of energy supply from sources such as wind and solar as sun could be impeded or wind might not blow. Hence, depending solely on renewable sources might add to the burden of government who may prefer already tested and reliable fossil fuels alternatives. With new technologies being develop daily such concerns are now being resolve through co-fire energy sources. There is the need to integrate fossil fuels and renewable energy mix through a process of cogeneration or co-fired power plants. The implication is Nigeria can push for the dual goal of renewables and fossil fuel concurrently.

Externally Driven Policy—Most policies on clean energy are fleshed out by international bureaucracy far remove from Nigeria's biogeophysical realities. Policies emanating from such institutions are always weight with external bias. An example of external bias is in the activities of the World Bank. The World Bank has not really helped matters as it is aiming for two contradictory objectives. The Bank is a major sponsor of fossil fuels globally while still promoting decarbonization through bioenergy alternatives. Different reports and figures indicate that the World Bank's funding for coal power stations have increased 40-fold in the last five years to hit 2.8 billion pounds in 2010.[11]

Similarly, the World Trade Organization (WTO) is an international body with the mandate of negotiating and regulating international trade. WTO remit is to promote free trade and stimulate economic growth. However, the WTO is criticized for not effectively implementing environmental policies as part of its activities on trade. As such the environmental impact of trade and labour movements is overlook. Rather, the environment has been inappropriately made use of by rich countries for trade protectionist purposes. In Khor's (2010) opinion, developed countries use this as a principle or concept to justify unilateral trade measures against the products of developing countries that do not have financial resources or access to low-emission technologies, and thus violate the principle of common but differentiated responsibilities.[12]

Maximizing Clean Energy Transition in the Green Economic Recovery Plan

Clean energy transition is viewed largely as a correlate of any workable framework in attaining sustainable development. This principle underscores strong linkages between poverty alleviation, environmental improvement and social equitability through sustainable economic growth. Clean energy benefits are not all about maintaining biodiversity and decarbonization alone as diverse forms of renewable energy can create employment, contribute to foreign exchange savings, and reduce dependence on imported sources of energy. Given that a transition to clean energy future would heighten skill while also creating employment Nigeria is developing its roadmap to a low carbon economy. Clean energy devolution is creating significant employment opportunities for different facets of the energy sector.

Regardless, the deployment of renewable energy is taking place at the same time with increases in fossil fuel energy consumption and production. For this reason, the latter is still the dominant sector. As such, transition to clean energy for low income countries is a complex issue with confluence of factors posing serious barriers towards building a green society. Clean energy transition sits on the crossroad of three major security challenges confronting Nigeria: energy security, food security and physical security (climate change). Now there is the fourth and most important dimension: existential threat of coronavirus.

In the aftermath of the COVID-19 crisis, carbon prices and carbon regulations which are already in place might find themselves under renewed attack. Governments must stimulate demand across multiple sectors of the economy. This will be more effective for economic recovery and have more lasting economic effects than deregulation. Although carbon prices and regulations might be perceived by some as a cost today, they are essential policy tools to internalize current externalities and incentivize the search for least-cost solutions to climate change, building a resilient economy and lessening the risks of major climate-related economic crises in the coming decades.

Pathways to Clean Energy Future in Post-COVID-19 Era

Various forms of clean energy are being deployed simultaneously across Africa. The energy sector remains the fastest growing sector of the economy. However, traditional biomass is still dominant sector as it is the primary energy source of most communities. In transforming the energy sector, Nigeria must turn the dream of clean energy development in to reality. The mechanism for releasing this dream is through integration of biomass into investment schemes capable of not only offering returns on investment but also having co-benefits for the host communities. Therefore, to fast track Nigeria's comatose clean energy sector requires biomass as building block for a sustainable energy future. Nevertheless, other sectors are equally important if the nation is to win the clean race without its negative effects. To fast track CET, Nigeria must carry out the following far-reaching transformations:

Climate-Smart Agriculture (CSA)—To carry out transformative change requires integrating the largest single employment providers into any plan of action for the future. This is because agriculture is a mega business with various forces involved in farming activities. Agriculture is the mainstay of the country with over 70% of the people engaged in agriculture related activities. These activities call for efficiency, output and profitability. Climate resilient agriculture foots this bill guaranteeing eco-efficiency while placing the continent on a pathway to sustained green growth. Climate-Smart Agriculture (CSA) is the building block for transformation of the agricultural sector. It seeks to improve productivity of livestock and farmlands rather than mere physical increases in cattle headcount and lands under cultivation. The major objective of CSA is to achieve green growth through ecologically sound principles of efficiency, productivity, mitigation, and adaptation. The Great Green Wall of the Sahara and the Sahel Initiative is a step in the right direction if the nation can harness its co-benefits of biofuel production. Though the environmental benefits in terms of reforestation and restoration of agro-ecological practices are immense, such transformation might not succeed without the benefits of sustained food and biofuel crop regeneration.

Clean Cities and Industries—Green economy creates an amiable environment for sustaining industrial, building and transport systems of most societies. Clean technology is the mechanism for removing waste in society by charting a pathway to efficiency and effectiveness in resource use. As a result, industries embarking on low carbon trajectories should be guided by ecologically sustainable practices with inter-temporal equity consideration as their goal. For instance, in the transport sector a low carbon transition should emphasize fuel efficiency, stricter standards for passenger and cargo transport, improvement in urban transit, and purchase of hybrid vehicles with high fuel economy. In Nigeria, Lagos State is championing this drive by enshrining clean transport pathways as part of their sustainable energy system.

Private Public Partnership—Investments in clean energy recorded impressive growth figures globally. Nigeria has not been totally left out of the clean revolution as clean energy investment is rising daily. Active participation of the IMF/World Bank propelled growth to new heights. Still, investments in clean energy are inadequate with vast renewable potential untapped. Government leadership is vital to channel investment into clean innovations and devolvement of the technologies. To jump start the process, a clean energy transition will require active private investment. In this regard, Small and Medium Enterprises (SMEs) as an important component of rural transformation must reorder their priorities towards providing critical clean energy services to communities. Because of the vast geographical spread of SMEs the devolvement of clean energy will then be socially inclusive.

Rebranding Biomass—As efficiency, productivity and environmental benefits underscore clean energy transition; crucial elements of people's livelihoods must also embrace energy efficiency for green society to take root. As biomass engenders climate risks, rebalancing livelihoods of the smallholder farmers such that they reduce waste and eco-costly inputs should be the stabilization goal of a clean energy future. To this end, biomass should be integrated into the national development plan with emphasis on priorities towards eco-sufficient alternatives. To this end, climate friendly technologies of solar panels and clean stoves proliferated. Food/energy crops contradictions can be resolved by proper land use management and ensuring land owners are also actively involved in all agricultural activities in their communities to ensure sustainability of their endeavours.

Power—Power is life as energy sustains life. All human economic activities have energy imperative as power underpins modern economic activities. Hence power or the absence of it determines the nature and form of clean energy technologies deployed and the cost of such green development. Therefore, extreme poverty is an offshoot of the absence of adequate power supply. Sadly, Nigeria's electricity consumption per capita is in rapid decline. Despite this, Nigerian energy supply of off grid solar energy is rapidly spreading. In the last few years over 1 million off grid Nigerians have replaced their kerosene lamps with solar lights. Even though it might be about 1% of off grid population it is on the scale to introduce a more efficient technology. With the price of photo-voltaic panels falling daily, Nigeria can tap into their vast solar power as the cost of gas and coal generated power are plummeting.

Similarly, swathes of land offer opportunities for wind, hydropower and geothermal energy generation. Nevertheless, fossil energy with fossil fuel subsidies is most attractive for the mere fact it is more constant. To stem massive fossil dependency, government can promote mixed sources of energy to tap in to the abundant energy potential on the continent. In particular, the hydro power sector is a promising area where only 7% of available capacity has been utilized. To limit fossil-based growth, government disincentives in the form of eco-taxes built into fossil production and consumption will make it less attractive to energy firms. The envisaged eco-tax should shift the burden from labour to end users and producers. This will create incentives for businesses to engage in renewable energy channelling sufficient investment towards carbon free innovations and technologies.

Nigeria: Now What Next?

There is no universal roadmap to a clean energy future, neither is there any certainty poverty and food insecurity will end by 2030 as the Sustainable Development Goals (SDGs) process seeks. This explains why despite clean energy ground breaking transformations in developed economies, Nigeria has not really reaped any tangible benefits from the green economy. Rather, most grandstanding proclamations on green economy only serve as impetus for further externalization of cost to the family farmer who bears the burden of all agro-ecological deteriorations.

Indeed, Nigeria's clean energy deployment has never really transcended issues of profit and shareholder returns in the first place. Though market driven approach is re-engineering energy consumption and production, but all is to achieve an end: Capturing virgin markets and new commodity frontiers. In this skewed relations, altruism and profit machination play complementary roles. But we know altruistic inclinations and profitability adjuncts are two strange bedfellows.

Hence Nigeria should adopt co-generation and self-generation from biomass, solar and wind to meet energy needs though this has disruptive implications for clean energy transition as some might inject surplus power generated into the local grid. Lack of adequate and consistent power generation is bane of power utilities that have not been mindful of consumer's requirements. To correct this requires formulating national clean energy strategies with the consumers as their goal. Such clean strategy alone does not ensure success of the endeavour. Political will and active participation of all stakeholders is critical to create efficient clean development. Stakeholders must embrace the culture of green devolution of technology for socially inclusive growth. Different Administrations are to blame as they have not totally embraced transparent clean energy transition in the unbundling of energy companies hence are enmesh in corruption charges. Establishing a regulatory framework governing involvement of green investors can protect livelihoods of the poor.

The flashpoint in the years to come will no longer be on climate risks but of how to redress the food crop and energy crop paradox as there is imminent threat of hunger in transitioning to clean energy as fallout of externality crisis. Traditional biomass must be re-engineered and this should form the building block of the nation's clean development. This will require scale-up of biomass to take care of energy market challenges. Still, such scale up must address the peculiar needs of the poor for food in the national blueprint.

Nigeria is at the bottom of the energy ladder. Thus, the quest to leapfrog on the heels of externally driven technologies is delusional in that endogenous technologies are likely to be the main growth drivers. So, Nigeria must build her technology base to confront the rising needs for technology. Skill is an important component in the shift to clean energy because skills determine the level of innovations in a country. Women are a constituent part of the clean transition and thus cannot be left out. Integrating women who sustain the family farmer's livelihoods into any future blueprint should be the overriding goal of government. However, the national roadmap should be based on the country's specificities and not the 'one-size-fits-all' approach envisioned by the international bureaucracy.

Regardless, Nigeria's clean energy future is now taking shape. The country has opened to clean business as various consumers are driven by the shortfall in energy generation to deploy diverse renewable energy technologies. Different communities are now embracing renewable energy without government's leadership role. This should not be. Government should play leadership role to ensure synergetic integration of various renewables into the national policy. To the government, nascent clean economy is fast gaining momentum. But in reality, it is on cruise control. Success in transition to clean energy is not going to be measured by what government value, think or say but by the availability of clean energy or how greener the economy is. For sure, success will be largely determined by whether captains of fossil industries allow green infrastructure propel the country towards green civilization.

Nonetheless, teething problem in the hydropower sector is manifesting with some projects posing serious constraints to government sustainable energy for all policy. The country's ability to meet the SDG #7 will be more realistic if they can tap the vast renewable potentials. The country's exploitation of the vast renewables tends to give different impression of the decarbonization initiative. It seems government has not given high preference to renewable energies as it deserves. Though renewable energy is catching on, the energy trajectory is not moving toward a cleaner pathway at the speed to reduce the impact of climate change while also improving access to efficient energy services.

To reduce emissions, a form of eco-tax introduced as disincentive. Disincentives in the form of eco-tax and eco-labelling built into fossil production and consumption will make it less attractive to the fossil fuels industry. The envisaged eco-tax should shift the burden from labour to end users and producers. This will create inventive for businesses to engage in diverse renewable energy thus channelling investment towards carbon free innovations and technologies. Smart-grids are critical for clean energy transition using diversification of the supply chain to curb challenges of intermittency, stranded assets, frequent power outages, lack of system capacity, high system losses, absence of proper grid infrastructure and quasi-centralization of energy generation and distribution. This requires a mix of energy pathways and scales that does not only synergistically integrates these complex issues but promotes favourable investment climate with prospect for improving energy access while simultaneously ensuring the energy transition.

The country embarked on a low carbon trajectory hopeful the abundant resources will ease risk of some of her fossil fuel infrastructures to be stranded assets. But the international oil spike of 2020 extensively brought to the fore shortcomings of Nigeria's projections. It changed the dynamics within and without the country since they all face health-related crisis. Hence, determining the type of renewables to be deploy where and when they should be deployed and their consequences for energy access should comprise the strategic end of the clean energy transition. This requires a new source of funds. It would require a new business model and management culture where risks of embarking on a new path are minimize while also ensuring energy access for all the people. Nigeria's renewable energy-based capacity is growing rapidly but from a very low base except for hydropower. Nigeria's energy transition involves significant additional policy interventions seeing that the vast fossil fuels infrastructures will be needed until 2080 when it will start decreasing either owing to exhaustion of resources or the new wave of technological shift overwhelm the system. To overcome wave of adverse dominion of nature, the implementers of the national policy should have a clear and undiluted view of the future.

Sustainable resource use requires synergy and trade-off of values balancing the best values from diverse assets to bring about equitable distribution of resources. The new roadmap for power sector reform should be implemented in so doing adopt best practices vital for energy efficiency and conservation in achieving the national decarbonization objective. To this end, crucial elements of the people's livelihoods must embrace the efficiency and sufficiency regime for energy transition to take root. Inter-sectoral synergies are needed to reduce the energy access gap given that onshore wind, biomass and hydropower are becoming more cost competitive than coal and gas fired power stations despite investment risk being higher in Nigeria than the global average. Rebalancing the livelihoods assets of smallholder farmers such that they reduce waste and eco-costly extraction should be one of the decarbonization goals in post-COVID future. To achieve this, biomass sector should be integrated into the national development plan with emphasis placed on priorities toward eco-sufficient alternatives such as bamboos.

The cultivation of bamboo and rattan on a commercial basis in the post-COVID-19 era foots government's energy access and energy transition bill by guaranteeing water, food and energy security while also earning additional foreign exchange. Bamboo is crucial for the transformation of the agricultural value chain in line with the Agricultural Transformation Agenda (ATA). Bamboo when efficiently cultivated in the Sambisa Forest will not only act as a food supplement, but it is with the co-benefit of generating additional revenue for the smallholder farmers while also improving environmental quality. The conflict-prone region of the Sambisa Forest and Lake Chad Basin in the North-East of Nigeria provide vast swathes of land to catalyse a green revolution. In the Lake Chad Basin, cultivation of bamboo serves the triple purpose of watershed management, reforestation purpose and income generation. In Sambisa forest, it will serve the purpose of improved environmental quality simultaneously dislodging Boko Haram extremists from their stronghold. This will also provide some of their followers' gainful employment and earn carbon credit for the country. Cultivation of bamboos would align national fuelwood consumption towards bamboos which is less carbon intensive while cashing in on the $65 billion international bamboo market.

Notes

1. Emefiele, (2020).
2. Matthew Green in Reuters of 5th May, 2020.
3. *Christian Aid (2019),* Counting the cost 2019: a year of climate breakdown.
4. *IPCC (2018),* Global warming of 1.5 °C.
5. *IRENA (2020),* Global Renewables Outlook, Energy Transformation 2050.
6. *Energy Transitions Commission (2018),* Mission Possible, How to reach net-zero carbon emissions from harder-to-abate sectors by mid-century.
7. Sachs and Tubiana, (2014). Pathways to Deep Decarbonization Pathways: 2014 Report.
8. Okoh, A. I. S. Pathways to Green Economy in post-Paris era Africa.
9. Tarheripour, F., Hertel, T.W., Tyner, W.E. (2009). Implications of biofuels boom for the global livestock industry: A computab'le general equilibrium analysis.
10. The Punch Newspaper, October 14th, 2014.
11. ChristianAid (2012). https://www.christianaid.org.uk/ActNow/climate-justice/worldbank/background.aspx
12. Khor, M. (2010). Challenges of the Green Economy Concept and Policies in the Context of Sustainable Development, Poverty and Equity. In The Transition to a Green Economy: Benefits, Challenges and Risks from a Sustainable Development Perspective.
13. Energy Transition Commission, 2020.

Chapter 9
Aligning the New NDC with NESP

> Pandemic is not a word to be use lightly or carelessly. It is a word that, if misused, can cause unreasonable fear, or unjustified acceptance that the fight is over, leading to unnecessary suffering and death. All countries need to review their Strategies now.
> Dr. Michael J. Ryan, WHO Informal Advisory Committee
> We all have to confront the fact there's not a magic bullet, short of a vaccine, that's going to make this [COVID-19] go away…We're going to be living with it. And we're not having that discussion at all.
> USA Today, May 11, 2020

Overview of the Extent NESP Aligns with NDC

Is the NESP roadmap aligned with the NDC in building a carbon neutral and energy just transition in the post-COVID-19 NDC submission? What are the potentials and barriers for integrating the NDC and NESP and how can the integration strategy be aligned with national development policies, programmes and strategies, designed to ensure sustainable development while also meeting the ambitious goals in Nigeria's new NDC submission of 2020? Are the institutional arrangements for different sectors consistent with the ultimate goal of Nigeria for reducing emission intensity of the economy while nurturing green growth and how can these diverse objectives best aligned with national development plan and the new NDC submission of the country? What are the viable options for building Nigeria's post-carbon society and curbing threats of climate change/COVID-19? This chapter will answer these difficult questions.

To start with, Nigeria acceded to carry out binding cuts simultaneously realizing the ambitious goal of becoming the 20th economy in the world. Tensions at multiple scales trail pursuing two inversely related goals of fostering fossil-led growth and meeting the Nationally Determined Contributions (NDC) targeted at reducing emissions. The 2020 COVID-19 crisis caused economic and health-related emergencies. Establishing the extent of implementation of the original NDC and its linkages with the Economic Recovery Growth Plan (ERGP) together with the new National

A. S. Okoh, *Oil Mortality in Post-Fossil Fuel Era Nigeria*,
https://doi.org/10.1007/978-3-030-60785-2_9

Economic Sustainability Plan (NESP) is crucial for the new NDC submission of Nigeria in the post-COVID-19 Age.

Nigeria is a party to the UNFCCC. All parties to the Paris Agreement are obligated to formulate and communicate short-term and long-term low-greenhouse gas emission development strategies (LT-LEDS). In particular, Article 4.19 of the Paris Agreement calls for Parties to "strive to formulate and communicate long-term low greenhouse gas emission development strategies" (LT-LEDS), and the Paris decision states that Parties "should strive" to communicate these "mid-century" LT-LEDS to the United Nations Framework Convention on Climate Change (UNFCCC) by 2020.[1]

Effectively, the Parties are required to communicate a new or updated Nationally Determined Contribution (NDC). Nigeria ratified the Paris Agreement in November, 2015. In line with the country's NDC, Nigeria developed and finalized Sectoral Action Plan (SAP) for implementation of the NDC in key priority sectors of energy, oil and gas, agriculture and land use, power, and transport.[2] Collectively, these sectors are significant to the economic development objective of the country.[3] However, these sectors vested with stimulating inclusive and non-inflationary growth are foremost sources of GHG emissions.[5]

Correspondently, Nigeria's long-term plan known as the Vision 2020 blueprint aim at growing the economy to become the 20th most developed nation in 2020. In line with the long-term goal, the Economic Recovery and Growth Plan (ERGP), a medium term roadmap targets Nigeria's vast natural resources to catalyze economic recovery.[6] The drivers of ERGP as enumerated by Preston Consult (2017) are: doubling tax compliance for non-oil sector; eliminating leakages at Customs; and reviewing rules for remitting surplus funds to FGN from independent revenue sources. The strategic thrust of ERGP is restoring growth (monetary and fiscal stability, external balance, economic growth and diversification.

However, the coronavirus disease of 2019 (COVID-19) changed all projections in the NDC by adding further burden to the economy causing a crash in price of crude oil. The implication for Nigerian monolithic economy is colossal. Although the country initiated policies targeted at economic recovery in the form of the National Economic Sustainability Plan (NESP) to stave off oil dependency while building structures to stem reoccurrence of the pandemic, Nigeria needed flesh into the NDC scale up of investment in both fossil-fuel and renewable energy infrastructure to foist near-term growth and long-term decarbonization commitments.

Nigeria Economic Sustainability Plan: An Overview

The 2020 COVID-19 has severe economic consequences for all economies- Nigerian economy inclusive. In Nigeria, the 2020 COVID-18 crisis negatively affected the economic fate of the country. COVID-19 crisis is both a health and economic emergency with rigorous corollary for the people and government of Nigeria. With surge of the virulent disease, public health infrastructure was the first victim of its devastation exposing Nigeria's deficiency in the sector. Nevertheless, the impact of

COVID-19 is not simply in the health sector only seeing that the pandemic also affected economic fortunes of the country. In the advent of COVID-19 pandemic, almost all global economies introduced different measures to stem the negative tide of the virus. A key approach adopted was the lockdown imposed to temporally suspend free movement of goods and people within and across borders. Effectively, the lockdowns froze economic activities causing massive job losses and supply chain disruptions. Unfortunate impact of the lockdown is that it depressed the demand for crude oil precipitating a crash of oil price. Most economies severely affected by the pandemic reduced or put a temporal hold on consumption of crude oil because transportation and production of goods aggravate community spread of the pandemic. As these sectors were caught in the vortex of COVID-19, global oil price crashed in April 2020 when crude oil sold at an all time low of -$10 in the US.

The ripple effect of this crash was felt across most economies. Nigerian economy heavily reliant on petrodollar for economic development felt the weight of the impact of the slump. Consequently, cracks in Nigerian economy between realizable incomes and projections at the beginning of 2020 turned into ditches. Filling the ditches became a policy nightmare. More so as the pandemic imposed crisis accelerated depletion of the country's dollar earnings in their reserves. The result of this weakening is depreciation of the Naira foisting upsurge in prices of good most especially those imported.

With a weakened average household purchasing power, capacity of the people to purchase items dependent on imported raw materials declined. Businesses that depended on importation for raw materials or other inputs for manufacturing were hampered by the sharp drop, by as much as 90% in foreign exchange earnings and shutdowns in exporting countries. The National Bureau of Statistics (NBS) projects economic growth could fall by as much as minus 4.40% to minus 8.91% depending on the length of the lockdown period. The situation is further aggravated by unemployment which stood at 23.1% (or 20.9 m people) in 2018 rose to 33.6 (or 39.4 m people) with COVID-19.[6] The economic impact of the coronavirus will be more acute within the large youth population who out of despair might engage in anti-social behaviour and criminal acts to make ends meet. With the recession resulting from the pandemic, government needed to take drastic measures to curtail the negative effects of the crisis.

In recognition of the impending crisis, His Excellency, President Muhammadu Buhari on March 30, 2020 established the Economic Sustainability Committee (ESC) to fashion a time-tested approach to tackle the challenges COVID-19 has posed. The Terms of Reference of the committee are [6]:

(i) Develop a clear Economic Sustainability Plan in response to challenges posed by the COVID-19 Pandemic;
(ii) Identify fiscal measures for enhancing distributable oil and gas revenue, increasing non-oil revenues and reducing non- essential spending, towards securing sufficient resources to fund the plan;
(iii) Propose monetary policy measures in support of the Plan;

(iv) Provide a Fiscal/Monetary Stimulus Package, including support to private businesses (with emphasis on strategic sectors most affected by the pandemic) and vulnerable segments of the population;
(v) Articulate specific measures to support the States and FCT;
(vi) Propose a clear-cut strategy to keep existing jobs and create opportunities for new ones; and
(vii) Identify measures that may require legislative support to deliver the Plan.

After consulting major stakeholders the National Economic Sustainability Plan (NESP) was presented and approved by the Federal Executive Council (FEC) in June, 2020. NESP is aligned with national economic plan such as ERGP and Vision 20:2020 and NDC. The NESP document required government to deploy a stimulus package, increase government spending, tax discounts, loan re-payment deferments or re-structuring, all with a view to increasing aggregate demand by beefing up investments and consumer spending. More, it is expected that if oil prices average $30 over the rest of the year, oil revenues (assuming Nigerian National Petroleum Corporation reduces Joint Venture operating costs by 20%), would amount to about N88.4 billion monthly. Assuming that non-oil revenues are sustained at the lower level projected in the revised budget estimates, the total allocations to Federation Account Allocation Committee (FAAC) for the rest of the year would then be around N485 billion a month. The roadmap further noted that this time last year total allocations to FAAC was N669.9bn monthly.[6] The very steep decline in revenues available for sharing among governments of the federation will have serious implications for wages, overheads and capital expenditures at Federal, State and Local Government levels.

Objectives of the NESP

NESP aims at mitigating effects of a deep recession, while ensuring social stability and addressing long-standing economic vulnerabilities envisaged in the Economic Recovery and Growth Plan (ERGP). The NESP blueprint has three main pillars: Real sector measures, fiscal and monetary measures and implementation measures. The real sector measure seeks to use a mix of project and policy approaches to create jobs in agriculture and agro-processing, food security, housing construction, renewable energy, infrastructure, manufacturing and the digital economy. The second measure seeks to maximize government revenue, optimize expenditure and enshrine a regime of prudence with an emphasis on achieving value for money.[6] The third measure is on implementation of the different components of NESP. Here, the blueprint speaks to issues of execution of specific projects and the responsibility for ensuring synergy between stakeholders.

The primary goal of the roadmap is internalization of growth which is based on the President's mantra of 'produce what we eat and eat what we produce'. Hence it is a deliberate strategy intended to rein in foreign exchange constraints experienced with the coronavirus. At its basic level, NESP seeks to minimise imports of goods and services relying on local technology and materials for growth. For instance, the plan envision local inputs such as limestone and rocks which is in abundant supply in the country will be used for road construction to reduce dependence on imported bitumen engendering additional burden on the overburdened foreign exchange earnings.

Agriculture is the bedrock of the NESP. The emphasis on agriculture owes to its capable of not only upturning the downward spiral but will gainfully engage the youth who are highly susceptible to uncertainties of COVID-19 impose economic meltdown. Increasing the local content in all economic activities is seen as a path towards innovation and development of the country's industrial base which took a serious bashing with the lockdowns. The lockdowns exposed the country's inadequacies in terms of responding with local resources to deal with the menace frontally. Other Objectives are[6]:

i. To stimulate the economy by preventing business collapse and ensuring liquidity;
ii. retain or create jobs using labour intensive methods in key areas like agriculture, facility maintenance, housing and direct labour interventions;
iii. undertake growth enhancing and job creating infrastructural investments in roads, bridges, solar power, and communications technologies;
iv. promote manufacturing and local production at all levels and advocate the use of Made in Nigeria goods and services, as a way of creating job opportunities, achieving self-sufficiency in critical sectors of our economy and curbing unnecessary demand for foreign exchange which might put pressure on the exchange rate; and
v. extend protection to the very poor and other vulnerable groups—including women and persons living with disabilities—through pro-poor spending.

Gaps in the Nigeria Economic Sustainability Plan

The NESP document is an expression of government's resolve to not only survive this emergency, but emerge from it thriving. NESP ultimate goal is to transform adversity into advantage. Instead of the prospect of 30 m unemployed Nigerians staring the country in the face, the document claims to have the capacity to put up to 30 m Nigerians to work in the short term. The main thing the document has achieved is it has regurgitated age-worn development clichés hence it is with insignificant consequences for the economic fortunes of the country. The problems that bedevilled the old development plans still trail this present one. It is merely a rewash and recycling of development clichés with inconsequential import for the unfolding crisis. In fact, it cannot put any significant dent on the state of affairs rather the economy will be on a free fall. With time market forces will do what it is known for. The all abiding faith in economic growth is premised on the trickle down approach. Economic growth recorded in the economy will create wealth for a select few at the onset but with time, will become an opportunity for all to grow in the long run (this assumption is line with the dictum 'with time all boats will sail with the tide').

But they have forgotten that poverty, pollution and growth overlap, combine and even clash in complex ways but their complementarities is seriously undermined by the replication of extreme poverty. The erudite theorist Wolfgang Sachs in commenting on market liberals' postulation of growth being mutually beneficial to both poor and rich affirms: in a closed space with finite resources the underconsumption of one party is the necessary condition for overconsumption of the other party. This view does not entirely hold for the environment in so far as their mantra of 'the more is wasted the bigger the common wealth becomes' subsists. Nigeria is an apostle of this dictum seeing that depletion of hydrocarbon deposits is intended to make the country great. Although ecological depletion is a necessary condition for the survival of man, the choice for reducing pollution may not be there in the first place. This is because other forces far removed from the immediate environment of the people have determined all socializations of nature. By privileging one class over the other is to place a barrier over the poor's actualization of their dreams. It will be ethically wrong to expect the poor to continue at a subsistent level of existence when the means to improve their well-being is readily available. Thus, it is an assumption which amplifies political elite's chicanery for nationalizing ecological degradations onto the poor.

Potentially, nationalizing ecological degradation will create manifold problems on both ends of the demographic divide, especially between the superrich and the poorest of the poor. The political elite might face the same constraints but they do have all the necessary conditions and fund to circumvent the system. In Nigeria, this proves to be a bane hence is another economic time bomb. Resolving these complexities of degrading for sustenance and polluting for convenience is the hallmark of the present Nigerian political dispensation.

Theoretical contradiction—NESP is a misnomer. The action plan is a hotchpotch of different theories conflated into one. The roadmap lacks conceptual rigidity in that it is aligned with differrent economic reorganization tenets. In one breath it seeks social sustainability, yet it is heavily dependent on exhaustion of natural capital. It failed to enshrine distributive justice in its outlook given the lackluster distribution of palliative. More the fact, implementation of palliative is left those who came as physicians but in real sense are the patients. NESP critics have argued that in limiting the perspectives the statement of intent has actually failed to concretely articulate manifold challenges of the economy instead points towards new BaU approach of the administration who are in their last term in office.

Ecological neutrality—In examining the ecological neutrality of NESP one will agree it is exhaustion intensive since it is based on the assumption of crude oil price stabilizing around $30 by the end of the year. The good news is Brent crude oil sold at $43.43 on 6th July, 2020. So, the economy can pull out of recession not because of the plans laid out but more the fact of the invisible hands of the market. Even then, it will require extensive pillage of petroleum resources to bail the country from the fiscal deficit. Accepted, the roadmap has the prosaic notion of rescuing the economy from its downward trajectory with Nigerians immediate economic needs as its focus. It is also compromising Nigeria's ecological integrity. And that is the ultimate price the country will pay since achieving eco-integrity, in a way that the needs of all people- today, and in the future- can be met and sustained is still uncertain.

Business-as-usual—Again, the blueprint is accused of being actively constructed to pry open new frontiers of rent seeking. Underpinning this is its conscious effort of steering economic activities towards infrastructural development; a pathway that will not only detract from inclusive growth but will only reinforce the status since the funds will end up where others like it in the past are: In private accounts of the lenders of the loan in Europe. In this sense, NESP only serves as an instrument for fundamentally reordering human relations. When viewed in this light, the ultimate goal of NESP is to ensure funds flow back to the country where the loans used in implementing the NESP agenda were derived.

Though this is farfetched, there is some elements of truth in the above assertion because corruption and rent seeking are intertwined with capital flight. It is, therefore, unthinkable to discuss one without the other. The basic problem, here, is not the motive of carrying out infrastructural development but rather is in the promotion of rent seeking activities which could exacerbate existing tension between the rich and poor in the country. The point here is not to do away with infrastructure's immense potentials of transforming the economy. To the contrary, it echoes its laudable tendency. The snag here is: the implementation entity is not totally free from the charges of corruption evident from the ICPC castigation of Federal Ministers for promoting corrupt practices.

Economic sustainability—It is old wine in new wine skin. The document has not only failed to exorcise the ghost of over labored Vision 20:2020 tenets that failed to take the country to promised economic El Dorado but reinforces the status quo with some crumbs thrown towards the common man as pro-poor policy. The blueprint is a 12 month plan of action to curb the downward trend of the economy returning it to a positive growth path. This is ludicrous. The timeline is not only too short but most of the implementation will spill into the oncoming government. A country beleaguered by policy inconsistencies will do well to steer clear of overlapping timeline.

Since the 2015, Nigeria's growth rate dropped to 2.7% with the economy contracting by an average rate of 2%. But 2020 COVID-2020 crisis changed the steady environment recorded in 2019. There is a fast growing labor force resulting from burgeoning unemployment rate of 23% in 2018 to over 40% in 2020 thus eroding the narrow gains in 2019. Contrary to expectations, NESP cannot lift the bottom half of the population out of poverty as growth trajectory intended in the roadmap is too low. This is allied weakness in the agriculture value chain. High inflation also adversely affects livelihoods of the urban poor.

Remember, poverty and environmental degradation is closely interrelated when formulating agricultural policies in developing countries. Nigeria's poorest class live in ecologically marginal areas and many of them farm on ecologically fragile land. Many of the poor farmers, who live in marginal areas, cultivate unsuitable land, thus exacerbating soil erosion, land degradation, and deforestation. Policies such as NESP can correct this if it adequately empower the poor explore their full potentials. As most have used up their seedlings in feeding their families it is difficult to reach the pre-COVID-19 farming activities. NESP targeted stimulus of 2.3 trillion will reduce the slide in the economy but cannot get the country out of the woods. In a country of over 200 m it is tokenism intended to create ambiance of rescuing the bottom half. In reality, government has succeeded in closing one gap and opened another more consequential malady; it has advertently or inadvertently contributed to the hardship experienced by the poor with the formulation of policies inimical to the economic well-being of the poor.

It may seem NESP placed an abiding faith in GDP growth hence the roadmap is concentrated on improving GDP growth. GDP is not a perfect reflection of economic reality of the country. Though GDP is the most commonly used index in developing economies but there is a problem: The problem with GDP is it not an all embracing toolkit. There is definitely need for a better measurement than it. Though economic growth cannot be measured in increased production alone, we must define a specific pathway to the future we want before defining ways of measuring how we are going to achieve it. This will then mean putting more pressure on government for more regulations on banks and money markets, for equity between rich and poor in the country, for investments that strengthen the poor, the unemployed, and the environment which is a sure way towards solving the problem. Breaking with present consumption will bring about an end to the miasma of economic problems faced in Nigeria.

Macroeconomic Environment—The weak macroeconomic framework of the country is another constraint. NESP roadmap will lead to high persistent inflation, multiple exchange rate windows and Forex restrictions. Correspondently, distortions by the central bank and lack of revenue-driven fiscal consolidation will be more prevalent as the economy will be in free fall. Rising public debt portfolio and complex policy interventions by the central bank will constrain private sector credit growth. This is more so as external balances are fragile to hot money movements, and fiscal buffers are exhausted, making Nigeria's economy vulnerable to external risks.

Another shortfall is the measure of expanding some sectors envisaged in the roadmap is weak despite massive fund committed to job creation with the view to stem the rising unemployment rate. Moreover, NESP roadmap did not include structural policy reforms. The country's medium-term growth is projected to remain stable around 2%. But with the economy growing more slowly than the population, the people living below the poverty line's standard of living will only worsen.

The key to a prosperous future is in the people. Blaming COVID-19 is just a distraction from those who are responsible. NESP is creating further divisions by blaming the pandemic and oil slump for the adequacies of economic diversification schemes initiated. The goal is to distract citizens from government' responsibility for the problem because their actions and inactions culminated in the impact of the slump felt. Government makes laws, collect taxes, and spend the citizens' money on pet projects and white elephant projects.

Set against a policy backdrop of ecological unequal exchange, the poor will continue to be at the receiving end of all economic and climate risks. The poor are most vulnerable to water stress, poor sanitation because they live closer to sources of pollution. They depend on traditional biomass—charcoal and wood- for fuel contributing to the numerous health hazards they now face. Because they are powerless to overcome shocks and extreme climate events, the poor's vulnerability is reflected in decreasing productivity with untold impacts on the environment. By privileging the rich, there are bound to be micro-level 'poverty traps' which can aggregate to the national level. By doing so, poor regions like the North east impacted by human-induced (Boko Haram insurgents) and natural disasters are unable to fully recover so that their capacity is reduced and their vulnerability heightened in the process making future disasters more likely.

Social Interventions—The Economic Sustainability reiterated government's plan to grow the economy in line with this administration's welfare centred agenda. This is against the backdrop of the National Bureau of Statistics report that 4 out of 10 households in Nigeria earn less than ₦137,000 per annum[6]. With high inflation, Nigerians living below the poverty line are expected to benefits from the palliatives in the NESP document. Unsurprisingly, job creation received serious attention in the policy toolkit. According to the NESP document it will provide 33.6% (or 39.4 million people) gainful employment in 12 months. The blueprint accurately predicts that economic growth could fall from −4.4 to −8.91%. It also presciently assumed that if oil revenues stay at $30 per barrel, the FAAC allocation for the rest of 2020 will be around ₦485 billion per month.

This is against the background of last year's average monthly FAAC allocation which was ₦669.9 billion. With unemployment at 33.6% the contracting of the economy becomes more significant. However, household income remains very low despite the palliatives and incentives to simulate the economy. Indeed, Nigeria has one of the highest poverty levels in the world. A stimulus package targeted at stemming negative growth must address cross-cutting issues of unemployment alongside waste in the economy linked the neo-patrimonial character of the state. Rather than merely increasing funds injected into the economy, a change in the orientation is required. This, again, is similar to the ERGP which made overtures to eradicating poverty with no tangible outcome. The recent arrest of the anti-corruption czar over corrupt practices is not only indicative that corruption is deep rooted but it is now fighting back.

To recover from the economic and health crises, the NESP document recommends different measures but was silent on a cultural shift. The document placed much faith in capital projects without structural transformation that creates conducive environment for growth. Hence these measures will only lessen the impacts of the pandemic but will not stop further re-emergence of other distortions to the economy. In fact, the measures will temporarily cushion the impact for a select few but will not end the impacts for a vast majority.

Again, NESP proposes to improve foreign exchange, reduce unemployment and drive sustainable growth. To achieve this, the document seeks to increase local production and consumption. The focus on local content is intended to drive innovation thereby equipping the country to face uncertainties such as the COVID-19 imposed restrictions on the movement of goods and services. This is sound technically, if it can rebuild the fledging educational institutions. With insignificant attention paid educational sector, the journey towards a truly home-grown technological base is an optical illusion. The plan to expand agricultural production and engage youths in farming and agro-allied jobs is another laudable plan that builds on existing policy of government. But with the implementation blind sight of the previous policy, NESP may be travelling on a road to perdition. The gap between fund released and number of beneficiaries, between government's policy thrust in the welfare schemes and expected liberal market outcome.

Notwithstanding, NESP targets a sizable number of Nigerians in its social safety net schemes but has no social register of the names of the people who are living below the poverty line. The end result is even when funds are released the real indigent Nigerians who need government assistance are not the real beneficiaries of the schemes. The real beneficiaries are either faceless Nigerians or members of the ruling party and their families. (This author was in a forum where a list was passed around for sharing palliatives between party members. An interesting issue then was that only party loyalists and their families were permitted to fill the form. The state government patrons were allotted slots with the remaining number shared amongst party hierarchy in the local government, then district level. Some ward level representatives of the ruling party who are closer in terms of economic means to the expected beneficiaries were left out). One would expect that as the impact of the pandemic cuts across party lines and within all sectors, the beneficiaries' should cut across party lines. Unfortunately, revise is the case as cronyism which waned in the beginning of this present administration is raising its ugly head.

The palliative should be cognizant of not just creating safety nets for the vulnerable class but should also address the temporal equity dimension between the present and future generation by ensuring the fund is used on items that can be handed down to the next generation.

Other issues of note are the handling of infrastructure development such as housing and power alongside providing support for businesses. Prospect of infrastructural development transforming the country is low as it will only transform the pockets of the superrich. Another issue is improvement in digital technology to increase learning thus increase access to education for school children in the COVID-19 age. Though some of the schemes introduced are positive responses to the crisis, there are concerns over implementation and accountability gaps. In the NESP document, the 2.3 trillion packages will be implemented by the Ministers within respective ministries where the schemes are housed. With the character of state accountability is Nigeria's albatross. It remains to be seen how a political class with short term vision will implement policies without considering their private ends. Digital education is sound technically speaking but with power supply still epileptic how can this project be achieved? Strict accountability cannot be assured if political office holders are the implementers of the schemes at the same time the accounting officer.

Policy inadequacy—The N2.3 trillion earmarked for economic recovery is inadequate. For one thing, the stimulus cannot meet the rising need of a population. Still the source of funding the projects is questionable giving that it is extensively reliant on foreign capital. Government should provide details of source of the N500m. This raises the question of the outcome of the recovery plan if the lenders fail to provide fund as expected and when expected? Another question that comes to mind is will this not add to the debt burden of the country which is overweighed by external borrowing to finance the budget? In an era when most economies are weighed down by coronavirus are there borrowers who are willing to release fund to others without posing harm to their economy?

The plan is sound theoretically just as it appears fantastic on paper. Even then, the aforementioned questions are not merely relevant but can undermine the agenda because the process of implementation is still questionable. In an economy where over 99% of generated revenue in the first quarter of 2020 went for debt servicing how do you implement a recovery plan without an alternative local source of capital? Undue reliance on petrodollar is Nigeria's Achilles heel. When issues such as payment of salaries, overheads, and project costs are extensively reliance on loans from external donors then the country is living in denier. The real value of Nigeria's stable economy today is the huge economic and ecological debts we are leaving behind for future generations.

Equally significant, is the treatment of other sectors. Other sectors of the economy such as hospitality, manufacturing and aviation were given insignificant attention thus living a huge hole to fill with these sectors pulling the economy into further depression. Increase in the poor's purchasing power intended to increase liquidity through palliatives will be eroded by rising prices of goods and services. Reflating the economy requires inclusive stimulus package targeted at all sectors with the prospect of strengthening the weakened Naira.

In the 2020 coronavirus crisis, government has an opportunity to accelerate the pace of structural reforms to build an institutional and policy framework capable of managing the cross-cutting issues of volatility of the oil sector and supporting sustained growth. Such transformation is a herculean task as it requires reforms that could have significant impact on the economy. Though there is now a gradual removal of subsidies, in reality the subsidy regime is only reacting to market forces. Elimination of subsidy is not totally in the hands of government but in the hands of party patrons who control policy formulated for the sector. Another reform is to eliminate forex and trade restrictions, greater transparency and predictability of monetary policy and increased domestic revenue mobilization. Such reforms would help raise the living standards of low-income groups while increasing spending on much needed public services.

Impediments to Synchronizing the NDC to NESP

Nigeria is grappling with different threats to its corporate wellbeing but two interrelated constraints have far reaching consequence for the ship of the Nigerian state. These issues are: Public health challenge of COVID-19/dwindling economic fortunes and inability to meet the upsurge of climate-related risks. These two cross-cutting issues have one thing in common: they are principally offshoot of natural factors, as well as human induced causation. The former impediment is the outcome of several interconnected issues, paramount amongst which is public health services failure, uptake of renewable energy technologies consequence of an equally important landmark event—the Paris Agreement of 2015. The latter constraint is a function of economic and environmental limitations also linked to the carbon intensity of the economy. Tackling these problems require filling up gaps created by these closely knit issues. Putting Nigeria's carbon-based economy back on the track of symbiotic relationship with nature is one of the solutions advanced at the international level. Other constraints include:

Institutional bias: To realize these goals require embracing new norms and values as well as new forms of technologies and climate support services capable of meeting emerging needs in a changing world. But the institutions to build the synergy in Nigeria are weak. Hence, there is a need for a study of the existing gaps with the view to creating awareness and strengthening skills of the different sectors for integration of NDC to NESP. More presently, institutions vested with implementation of these policies treat the NESP and NDC as mutually exclusive, hence are policies standalone items without conscious efforts of synchronizing these objectives into a holistic framework.

Carbon-based growth: Nigeria's economic transformation requires burning dead or living carbon. Burning dead carbon fuelled extensive discoveries that resulted in remarkable improvements in living conditions globally. Carbon emission is attractive to most developing economies because energy is an integral part of economic development. In a developing economy like Nigeria, burning dead carbon provides households and industries with much needed energy. Energy generation is, therefore, crucial to national wealth creation, employment and improved standard of living. Nigeria keyed into this free market assumption of carbon-led growth being a necessary precondition for the transformation to a middle income nation. An explanation for the Vision 20:2020 blueprint, which intends to optimize human and natural resource potential to achieve rapid and sustained economic growth and translate economic growth into equitable social development that guarantees a dignified and meaningful existence for all citizens,[7] but uses abundant fossil fuel energy. The simple argument is maintaining pace of economic growth at 6–9% annual GDP (Gross Domestic Products) growth will require increases in the extraction of natural capital to pave way for the transition to a high middle income country by 2020.

Economic growth imperative: Nigeria's fossil dependent economy recorded impressive rise in GDP before the economic recession of 2015.[7] It is trite to state that the remarkable transformation recorded in the last decade is the outcome of GHG emission. But that is the truth. In addition to biomass extraction, GHG emissions from gas flaring create an adverse carbon upload. Mitigation activities crafted at United Nations Framework Convention on Climate Change (UNFCCC) insists that different nations embark on a less carbon intensive pathway, requiring further squeeze on fossil fuels. Placing a squeeze on fossil production is, to put it mildly, a big task in Nigeria. The general sentiment within Nigeria has been one of feeding the citizens which is a bigger priority for government than focusing on the shift to clean energy, even if the benefits, and consequences of not doing so, are apparent. This captures Nigeria's sentiment encapsulated in the National Report to Rio+20 in 2012.[8]

Lack of political will: Growing trend towards economization of nature in resolving Nigeria's ecological logjam poses new concerns. Gas flaring serves a political end in a polity highly reliant on fossil fuel.[7] Over the years, different administrations have engaged this perplexing issue but with limited success. The non-committal response to this issue is fuelled by the political visibility of large oil firms. The presence of these firms at the political arena has not really helped matters. On close inspection, it could be linked to the tendency of seeing the petroleum sector as 'cash cow' for generating foreign exchange earnings. Government's argument has been one that the revenue from the oil sector will drive economic growth; since short-run benefits of gas flaring outweigh any perceived long-term environmental threats.

On the back of fossil driven development, the economy recorded impressive growth in the last ten (10) years. Nigeria's growth evident from the 2014 GDP rebase was, however, exogenously driven and heavily reliant on a single resource: the oil sector. Growth of the economy with extraneous determinant carries with it seeds of despoliation of nature. The carbon intense trajectory of Nigerian economy is linked to what Evans calls the 'Triple Alliance'.[9] Degradation is, however, not limited to the fossil energy sectors alone. Within the biomass sector, degradation exists- an outcrop of being an agrarian society. Even though government over the years enacted laws proscribing bush burning and deforestation, they yielded intangible benefits. The failure of applying stringent laws on this sector is directly attributable to lack of political will for providing alternative energy to use for domestic purposes in most rural communities. Furthermore, enforcement of the law might criminalize the poor for providing the basic energy required for their family. Allied to this, is the growing influence of timber prospectors within rural communities, which somehow dovetails into clientelists' relationship with political patrons at the national level. On this ground, addressing deforestation carries serious weight capable of undermining the political fortunes of the party in government. And as most politicians have short-term horizon, policies formulated only serve their immediate goal.

All said, on paper NESP blueprint sounds good and shows commitment towards bailing the economy from the potential pitfall. But the unfortunate truth is that NESP in its current form will not meet economic aspirations of Nigerians. For one thing, it is not only outside the prevailing economic circumstance of the country since most of the goals are not attainable. No doubt, some of the issues will be addressed within the prevailing circumstance of the country, other are still very fuzzy given the likelihood of the social compass to also encourage the re-emergence of corrupt practices at national levels. With this influence of captains of industry on the rise, a new generation of 'COVID-19 billionaires' who will extensively benefit from the new opportunities created by the recovery businesses will emerge. When there is an optimisation of business (profit) not all participants in the economic recovery scheme can fully benefit from the initiative. Already echoes of divergence are emerging from unlikely quarters.

Within business community, they are already crying out. On the 6th July, 2020 the Lagos chamber of commerce cried out of the non-inclusivity of the NESP roadmap. To them, the blueprint is not an inclusive agenda. So, if the economy must revert to its pre-COVID-19 steady growth, then the economic stimulus package takes into account their interest. This is the objective truth. Business and industry have a crucial role in delivering the economically viable products, processes, services, and solutions required for the transition to economic nirvana. Hence must be carried along. It seems they are crying wolf hopeful that their non-inclusion in the scheme never materialises. The real losers are Small and medium-sized Entreprises (SMEs) who are the real backbone of the economy accounting 48% of the national GDP. These SMEs mostly do not have unions, or bank accounts. Fears have been expressed—probably well deserved- in some quarters of the NESP not taking care of SMEs.

Premised on the implementation lacuna, one is forced to agree with its critics that it is not a royal road to the promised economic nirvana in the post-COVID age. Be that as it may, the blueprint is not cast in stone; it is still open to further clarification from the Federal Ministry of Budget and Planning. Till then, what we have as NESP tantamount to bailing out few buckets of water as the Titanic sinks. Yes, it is only a few bucket that will be taken from the sinking ship. This will cushion the impact of COVID-19 for some select few but is not be inclusive neither can it usher in a green growth trajectory.

For the economy to be on a true recovery trajectory, all the constituent parts of the economy must be carried along. This entails going beyond the narrow political terrain to distribution of the stimulus package to all facets of the economy with accountability as its watchword. What we have now as social safety nets providing palliatives that end up in politicians' pockets which is not justifiable. The unfortunate fact of this vagrant abuse is that the poor who are at the receiving end of all economic risks will still bear the brunt of all economic shortfalls.

Analysis of Aligning the NDC and NESP

As a party to the UNFCCC Nigeria acceded to make an unconditional contribution of 20%, which is below business-as-usual (BAU), consistent with current development trends and government policy priorities but will make a 45% conditional reduction with additional adaptation fund.[10] Estimated per capita emission using the LEAP scenario and Re-based GDP shows that the current emission is around 2 tons CO_2e 2030 with BAU it is around 3.4 tons CO_2e 2030, while conditional is around 2 tons CO_2e 0.873 kg CO_2e (2015) 0.491 kg CO_2e (2030) (Table 9.1).

However, putting an end to GHG emissions from fossil fuel energy in the NDC is with severe consequences for Nigeria's economic wellbeing as the economy is susceptible to external shocks witnessed in the 2015 and 2020 oil market downturns. Primarily, GHG emission in Nigeria is as follows: AFOLU- 66%, energy- 28.2%, waste- 3.0%, IPPU- 1.9% (FGN, TNC, 2020). Without ambitious mitigation action, within 45% unconditional based on LEAP scenario Nigeria's per capita emission is projected to grow to 3.4 tons CO_2e (Table 9.1). Agriculture is the largest sector and employer of labour. It contributed 23% of the GDP in 2013 and employed approximately 70% of the country's labour force. On average, the agricultural sector

Table 9.1 Key data on NDC

Information	Value	Source
Historical emissions (1850–2010)	2,564.02 million tonnes (MT)	CAIT database, World Resources Institute
1990 emissions	163.91 MT	Nigeria's Second National Communication
2000 emissions	214.21 MT	Nigeria's Second National Communication
2010 emissions	263.0	MT Energy Commission of Nigeria
Estimated emissions per capita	Current: around 2 tonnes CO_2e 2030 BAU: around 3.4 tonnes CO_2e 2030 Conditional: around 2 tonnes CO_2e	ECN (estimated 2015 emissions), World Bank (population estimate), LEAP scenario
Emissions per US$ (real) GDP	0.873 kg CO_2e (2015) 0.491 kg CO_2e (2030)	ECN, Re-based GDP LEAP 2030 ambition scenario
Emissions as % of global total	< 1% (2010)	ECN (2010 emissions estimate) and US EPA global estimate
GDP per capita (US$)	2,950 (2014)	Re-based GDP
	3,964 (2030; real 2015 US$)	LEAP 2030 BAU scenario

Source[10]

contributed 40.3% of GDP between 1999 and 2011. It rose from 36.7% in 1999, peaked at 43.9% in 2000 and stabilized at 40.2% in 2011.[2]

Given the need for development, emissions from conditional contribution are expected to stabilize slightly above today's level of 2 tons per capita, once the full package of policies and measures with the NDC has been implemented. Climate finance is the bedrock of the decarbonization, if the country will meet their target. But without international support unsustainable practices will increase given that biomass consumption is expected to triple in 2030 tallying with demographic pressures (Table 9.2).

Without ambitious mitigation action, within 45%, unconditional LEAP scenario Nigeria's per capita emission is projected to grow to 3.4 tons CO_2e (Table 9.1). But, given the need for development, emissions from conditional contribution are

Table 9.2 Demographic assumptions

Parameter	2015	2020	2025	2030	2035	2040
Population growth rate	3.52	3.80	4.00	4.00	3.88	3.74
Number of persons per household	5.60	5.60	5.40	5.20	5.00	4.00
Number of households	3.7	42.1	53.2	67.3	84.8	101.9

Source[11]

expected to stabilize slightly above today's level of 2 tons per capita, once the full package of policies and measures with the NDC has been implemented. Climate finance is bedrock of the decarbonization, if the country will meet their target.

The imperative to mainstream the NDC into both the new recovery plan and National Development Plan is high. A reason is the severity of socioeconomic challenges experienced with the COVID-19 pandemic. COVID-19 pandemic informed formulation of NESP. The primary goal of the NESP is job creation and infusion of cash. To achieve this investments in agriculture and agro-processing, manufacturing, renewable energy, housing, information technology has to be scale-up. Ensuring local content is a remarkable development given that it will act as a catalyst of innovation. Despite the good intention and the massive injection of capital in the economy, NESP sounds like the doomed Vision 2020 which is destined for the backwoods of history. The Vision 20:2020 hence ERGP plan sought to make the country a top 20 economy by 2020 through an annual GDP of at least \$900b and a GDP per capita of \$400.[12] But this is not tenable. Like the ERGP, the NESP stimulus package of N2.3 trillion funded by N500bn from Special Accounts, N1.11 trillion of CBN structured lending and N302.9bn from other funding sources.[12] Additional burden is placed on crude oil export to bail out from stringent loans. Furthermore, delegation of responsibility to Ministers of respective ministry invokes ghost of Vision 20:2020 without concretely exorcising ghost of bureaucratic short sightedness of political office holders.

Further, NESP proposes a Solar Home Systems project that will cover up to 5 million households, serving about 25 million individual Nigerians who are currently not connected to the National Grid. Due to scale of the materials required, solar equipment are to be manufactured locally thus creating additional jobs and growing the technological base of the country. Environmentally, co-benefits of this project include meeting NDC while also improving environmental quality. However, this is reminiscent of the contract for N9.2 billion worth of cooking stoves and wonder bags for rural women under the National Clean Cooking Scheme awarded to a South African firm for the manufacture and training of local artisans. This is sound reasoning as biomass consumption which is the primary energy supply of over 80% of the country[13] would have drastically declined. Alas, the decarbonization goal of government does not have strong political support for limiting emission at all levels. The want of local anchorage across all sectors undermined ERGP and this same constrain can also impede political lock-in of the NESP. This is with grave consequence for realization of GHG cuts in the new NDC.

The 2020 COVID-19 crisis could slow down Nigeria's effort towards meeting its long-term climate strategy. To avoid carbon lock-in (as natural gas will be the cheapest source of energy) requires aligning long-term goal with short-term NESP if the country will eliminate the risk of stranded assets. Sustained introduction of renewable energy in the SE4ALL 2030 plan will reduce risk of lock-in but reinforce fossil fuel asset abandonment crisis. If Nigeria implements all its Low Carbon Development (LCD) opportunities, GHG emissions for the BaU Scenario will decrease from the BaU level of 967.2 Gg CO2-eq to an LCD level of 797.2 Gg CO2-eq (Table 9.3).[14] The potential amounts to 170,016.5 Gg CO2-eq in 2035, representing a decrease of 18% on the BaU scenario emissions.

Table 9.3 Mitigation potential of the interventions by sector and category

Sector and category	Mitigation potential (Gg CO2-eq)
National	1,70,016.51
Energy	60,002.18
Fuel combustion	53,426.80
Energy industries	45,248.40
Electricity generation	45,248.40
Manufacturing industries and construction	405.3
Transport	4,897.90
Road transportation	4,897.80
Other sectors	2,875.20
Commercial/Institutional	−3,196.80
Residential	6,072.00
Fugitive emissions (Oil & Natural Gas)	6,575.38
AFOLU	1,10,014.33
Land	1,10,014.33

Source[1]

To limit fossil fuel assets abandonment crisis, structural transformation of both demand and supply side energy sector together with improvements in renewable energy generation will ratchet-up power generation but cannot end GHG emissions. For the country to curb GHG emissions, unconditional emission based on 13% using LEAP scenario (Table 9.4) will grow the economy but generates additional carbon. Projected energy demand for Nigeria (Table 9.4) based on a 10% GDP growth rate using NECAL2050 modelling tool indicates that by 2020 the total electricity demand in 2020, under a 10% reference growth will be 63,363 MW.

An ideal situation is 13% optimistic growth which enables the country meets her developmental goals. Under this scenario, 45% conditional mitigation will foster energy security using funds injected into the economy for LCD. This entails coordinated processes through shared institutional arrangements, common systems for monitoring and accessing progress that will enhance alignment of the diverse goals.

Table 9.4 Electricity demand projections per scenario, MW

Scenario	2015	2020	2025	2030	2035	2040
Reference growth (7%)	24,380	45,490	79,798	115,674	161,651	213,122
High growth (10%)	30,236	63,363	103,859	196,875	333,444	487,510
Optimistic growth (13%)	41,133	88,282	170,901	315,113	573,289	979,326

Source[15]

Bridging the NDC and NESP Fissure

The year 2020 and the 2020s decade hold great promise to the world community for not only being the start date for implementing new measures, the year is also the beginning of the decade for reaching overarching global goals by 2030. In recognition of it critical to world community, 2020 has been dubbed the *super year* by UNEP. The year is also when Nigeria's vision 20:2020 and ERGP will terminate. Hence need for another plan for the economic recovery in the post-COVID-19 era and another long-term plan. Furthermore, 2020 is a year when review and follow up actions on the Paris Agreement and Sustainable Development Goals (SDGs) are carried out. So, major decisions on post-2020 global frameworks will be taken in this year.

With different reviews and follow-up actions on the table, policy must be up to the task at hand if it must be relevant. The issues to be considered this year are not only cross-cutting but are mutually dependent. Therefore, policy needs to be fast and effective to cope with the scale of activities planned for the post-COVID-19 era. To be effective a combination of a suite of approaches including the 'carrot and stick', 'command and control', technology-push and market-pull instruments are necessary components of the policy space.

Synergistic integration of these instruments will change both the energy demand and the supply side of the power generation sector. Stimulating technology development is therefore necessary. Funding of research and development will create a technology-push fostering innovation in alternative technology. A technology development plan for long term deployment of new technologies will not only scale-up technology but foist innovation within the economy. Building codes and product standard in Nigeria are not targeted at ensuring efficiency. Technology standards set minimum requirements on energy efficiency, fuel economy, and emissions limits for vehicles and power plants. Though decarbonization is gaining limited traction nationally the efficiency gains have not been replicated in building codes and technology standards.

The PIGB strives to reduce gas flaring by enshrining paths oil firms must adopt to reduce burning of gases. Nevertheless, the Bill is enmeshed in several interrelated crisis. For one thing, Nigeria lacks policy similar to India's National Clean Air programme with emission-control standards on coal-power plants neither policy similar to South Korea's countermeasures after declaring air pollution a 'social disaster' exist. Nigeria need formulate policy similar to China's 5-year plans and the EU's Energy Efficiency Directive programmes targeting productivity gains in building, transport, and industry sectors.

To bridge the gap between NDC and NESP, the market is important to ensure that the long-term development policies of government are aligned with the NDC. Market activation through market-pull policies will promote market deployment of solutions and accelerate uptake to help viable technologies achieve a decline in unit costs. The DNV GL ETO of 2020 is of the opinion that this happens through learning-by-using and feedback for further technology development, industrial efficiencies, ongoing market-focused R&D, and economies of scale. DNV GL ETO suggests economic instruments such as tax reductions for renewable, subsidies for EV purchases and charge points, battery storage and feed-in tariffs for renewable power, and other influences on energy prices will stimulate market uptake.

Nigeria has recorded a myriad of challenges associated with market distortions. Economic signals to fix market distortions necessitate removing the barriers to clean energy transition as illuminated in another part of this book. A major bane is policy inadequacy of not capturing carbon price and unending fossil-fuel subsidies which weighed down the country. Nigeria expends one-third of its foreign exchange earnings on subsidy of an item produced in the country. Ratcheting up the NDC ambitions is needed if the country must meet the Paris Agreement. The main aim of the Agreement is keeping the global temperature rise "well below 2 °C". The country's commitment of 20% unconditional mitigation is the baseline for measuring progress in implementation of the Agreement. But this can only given effect if the country's climate ambitions for reducing emissions considers its domestic capabilities and circumstances imposed by COVID-19 reduced near-term economic constraints while promoting long-term decarbonization. But 45% conditional mitigation will reduce carbonization while also tackling uncertainties of exiting fuels concurrently building a climate resilient society equipped to face the vagaries of climate change and COVID-19.

For now, Nigeria's planned fossil-fuel production surpasses those production levels that are consistent with implementation of the stated commitment in the NDC. Effectively, the country's current NDC has insufficient ambition levels to reduce upload of GHG as minimal policy attention on limiting fossil-fuel production is paid by the government in the NESP which will hinder the country meeting its part of the collective efforts to meet the Paris Agreement goal.

The Way Forward

Consumption and production of fossil fuel is critical to NDC implementation and meeting developmental goals of government. Tensions between competing needs is commonplace undermining long-term decarbonization commitment. The NDC has received serious attention though near-term risks have railroaded government policy towards economic and health-related commitments. Different forms of transformations are recorded in the NDC but cannot trigger a low-carbon revolution or foster an inclusive green growth. However, COVID-19 short-circuited government's goal for the people and the environment. Though the environment is a major short-term gainer, the environment is a long-term loser since hydrocarbon deposits will fuel growth.

Meeting Nigeria's development plans simultaneously reducing carbon intensity of the economy, policy need play an important role. Market-pull policies will engender market deployment of solutions and accelerate uptake to help viable technologies achieve a decline in unit costs. Economic instruments such as tax reductions, subsidies for Fossil-fuel should be removed while eco-label and eco-tax and other influences on carbon up take embarked upon to stimulate LCD developed to improve power-generation capacity. Pricing carbon and other negative externalities (air pollution, environmental damage) provides a clear market signal. A carbon-pricing scheme can impose a tax on emissions or set a cap on emissions while allowing trading to achieve the most cost-effective reduction.

A rise in NDC ambition levels is critical to meeting the country's decarbonization agenda. How the country's revenue is spent is vital for political feasibility and public acceptability. Creating a Just Transition Taskforce ensure that revenue from CO_2 tax will be recycled and returned to the population on a per capita basis. An Emissions Trading System (ETS) will compensate households with a 'Climate Credit' on utility bills, and some ETS revenue go to a new fund dedicated to greenhouse gas (GHG) reduction known as the Reduction Fund. Hence social processes and institutional mechanisms must aim for concerns society value most. Moreover, identification of resilience-enhancing actions such as poverty alleviation scheme of N-Power, Trademoni and Marketmoni together with Anchor Borrower Scheme and Fadama III programmes can meet both developmental and environmental concerns concurrently if channelled towards low carbon priorities.

Others are carbon emission reduction mechanisms such as $REDD^+$ with the co-benefits of achieving carbon sequestration while raising additional income for host communities via carbon credit. The country should develop institutional structures effective in resolving the low carbon growth and economic growth dilemma. Such institutional structures is reinforce with a clearly articulated Low Carbon Growth Plan (LCGP) which lays out the direction and vision of development whereby contradictions of growing the economy while also ensuring decarbonization is resolve.

Nigeria's renewable energy drive needs policy levers to catalyze transformative change. Transforming these resources require international support from various financing mechanisms and leveraging Green Climate Fund, Adaptation Fund, Reducing Emissions from Deforestation and Forest Degradation ($REDD^+$), Forest Carbon Partnership Facility (FCPF), National Appropriate Mitigation Actions (NAMAs) to reduce fiscal pressures and attract private sector participation in sustaining low carbon development.

Notes

1. UNFCCC (2020), *Communication of long-term strategies*, UNFCCC Secretariat, Retrieved June 25, 2020 from https://unfccc.int/process/the-paris-agreement/long-term-strategies.
2. Federal Government of Nigeria Third National Communication (FGN TNC), (2020). *Nigeria's Third National Communication under the United Nations Framework Convention on Climate Change.* Abuja, Nigeria: Federal Ministry of Environment.
3. Federal Government of Nigeria, First Biennial Update Report [FN BUR1]. (2018).
4. Cervigni, R., Rogers, J. A., & Henrion, M. (2013). *Low-Carbon Development Opportunities for Nigeria.* Washington, DC: The World Bank.

5. Okoh, A. I. (2014). Green Economy Versus Green Growth: A Quest for an Ecologically Sustainable Polity in Nigeria. *SCSR Journal of Development, 1*(4), 22–40.
6. Federal Government of Nigeria (FGN) 2020, Bouncing Back: Nigeria Economic Sustainability Plan.
7. Okoh, A. I. S., Mulumo, D., & Orokpo, O. F. (2018). Assessment of the implementation of Nigeria's Nationally Determined Contribution (NDC) in transition to a Low Carbon Economy. *Nigerian Journal of Management Sciences, 6*(1), 372–383.
8. Federal Government of Nigeria (FGN). (2012, November 8). Climate Change and Sustainable Development in Nigeria. National Report for COP18/cmp.
9. Ikein, A. A. (1990). *The Impact of Oil on a Developing Country: The Case of Nigeria.* Ibadan: Evans Publishing.
10. Federal Government of Nigeria Intended Nationally Determined Contribution (FGN INDC). (2015). *Approved Nigeria's INDC Submission.* Retrieved from www.unfccc.int/submisions/INDC/submissionpages/submissions.aspx.
11. Bala, E. J. (2014). *Nigeria's Power Sector Reform: What Next after Privatization.* Johannesburg: Power and Electricity World Africa Conference.
12. Federal Government of Nigeria (FGN). (2017). *Economic Recovery and Growth Plan 20,172,020.* Abuja, Nigeria: Federal Ministry of Budget and Planning.
13. Okoh, A. I. S. (2017). *Pathways to a Green Economy in Post-Paris Agreement Africa.* Oxford: Green Economics Publishing House. Retrieved from https://www.greeneconomicsinstitutetrust.org/books/pathways-green-economy-postparis-africa/.
14. ECN - Energy Commission of Nigeria. (2012). *Electricity Demand Projections.* Abuja: Energy Commission of Nigeria.
15. Okoh, A. I. S. (2018). An analysis of Nigeria's nationally determined contribution (NDC) in the transition to a low carbon economy. *Open Journal of Social Science and Humanities (OJSSH).*[Online] 1 (1), xx-yy. Available: https://www.openjournalsnigeria.org.ng/OJSSH/08201801116.php

Chapter 10
Beyond Oil Age

'All in All, I Wish We Had Discovered Water'.
Ahmed Zaki Yamani, Saudi Arabia Minister of Petroleum.

The Way Ahead

We began this journey with a discussion of dawn of fossil fuel-led growth and its implication for the global economy. In the preceding chapters we showed how COVID-19 is undermining all projections and plans to reduce the carbon intensity of the economy. Building health infrastructure to sustain live while also ensuring industries are aligned towards net zero emission is the defining challenge of the post-COVID-19 world. Overcoming uncertainties of the coronavirus at the same time with enshrining a culture of attentive to nature is the new normal. In the journey we also showed not just the path but also the major achievements and shortcomings of building a durable bridge to a carbon–neutral future by exposing pits and ditches capable of derailing the clean energy transition. Nigeria is trying to resurrect the ghost of her petroleum-led economy despite redirection of the global economy towards new alternatives in post-Paris era. The PIB in its present form is not a long-term measure but a stop gap policy with a short economic vision. In promoting the PIB instead of laying out country specific low carbon and growth plan, Nigeria stands the risk of losing the advantage of being one of the forerunners in the renewable energy business.

In the preceding chapters we also brought out how renewables are economically and politically ideal for countries heavily reliant on imports to meet their energy requirements and as such the oil sector is destined for a moribund life in the coming years. Also, we saw how the Eureka moment is coming with the lowering of PVs price (60%) and decline in wind turbines prices (29%). But the brief euphoria of curbing fossil fuel might be eroded by COVID-19. This can either reinvigorate fossil fuels production or place the world on a part of symbiotic relationship with nature. It might

A. S. Okoh, *Oil Mortality in Post-Fossil Fuel Era Nigeria*,
https://doi.org/10.1007/978-3-030-60785-2_10

seem the latter is gaining more traction seeing that mature economies are professing green economic recovery tenets. Some of the co-benefits of a low carbon development are vividly brought out in the previous chapters including but not limited: energy efficiency, food security, reduction in dependence on oil, improved environmental quality, increased national security and reduced carbon footprint of the global economy.

In the foregoing chapters we further show how the low carbon transition would have manifold consequences for economic development thus hampering attainment of the Vision 20:2020, ERGP and other new policies of government. A worst-case scenario is that the energy revolution could erode gains of decarbonisation when insufficient fund resulting from the looming danger in the form of end of Oil Age stifles government's capacity to meet national commitments. Though some within government circles are avid supporters of diversification using agriculture as the springboard, others are diehard supporters of fossil fuels hence are holding on to the vestige of its past glory. Nevertheless, there is a strong inclination within the present government to pursue agricultural development but forces within the ruling elites have hijacked power hence are more predisposed towards capital expenditure for obvious reasons.

Despite government's commitment towards reducing reliance on fossil fuel, the country has never really wavered from her pursuit of this goal. The only sour spot in this quest is the long-winded approach of resolving problem of decarbonization to a level that some argued it borders on political rhetoric. Arguably, Nigeria's economic diversification has not really transcended rhetorical obligation which explains the lack of meaningful progress. To such critic, diversification is a catchword deplored as the leading refrain of the siren-song for national economic revival with occasional flashes of agricultural transformation occurring intermittently. What could have been a remarkable display of commitment is the Anchor Borrower's Programme but some claim it is still suffering some teething problems- a criticism which lacks merit for the mere reason that instances of serious abuses have not been uncover within the scheme. What one can say to be wrong is the progress made is marred by insufficient funds to drive diversification even though government's investment and commitment to agriculture has improved significantly. If additional funds are injected into this sector, uncertainties associated with the looming danger would be limited. Prospects of eliminating oil seem real but since there is no real progress in fashioning a strategy, this opportunity might be totally lost again. Overcoming climate risk is equally bleak given that the scourge is high, fossil fuels elimination is not catching up nationally at the rate required. So, the country must change its ways by ensuring that the political rhetoric of Paris translates into concrete climate actions of today. But this must be mindful of COVID-19 threat which is still looming in the background.

With the prospect of waning in spread of COVID-19 high (due to the discovery of a vaccine by Pfizer in November 2020), change is in the air. A sense of urgency and anxiety exist over the issue at global level: a *déjà vu* situation. And this is understandable. Within Nigeria government's anxiety is inevitable given the pace of spread of the pandemic together with economic recession. Unfortunately, the people on whose turfs these existential battles are fought have no say in the resolution of the impending catastrophe. When the decision on curbing fossil fuel is taken the major constituency, whose voice is going be recognized is only the fossil fuel industry and its army of lobbyists thus, ensuring the future gets further alienated from the people. The poorest of the poor who are at the receiving end of this onslaught from both man and nature have always known that when environmental goal and economic gains compete for attention, the latter always wins since it is aligned with interest of the elite who control instruments of governance. Even though technological advancements now may reduce spread of coronavirus and reduce adverse carbonization, the country is not deploying new technology as it ought to within health services/industries or innovative to change how we relate with market forces in accelerating phase out of fossil fuels. The downside of this is, it could exacerbate growing divide between the haves and have-nots. The net effect of this is some poorest of the poor in the country today would be worse off than they were in the past while the 'tribe of rent-seekers' with their acolytes will thrive on these distortions they have created within the system.

Nigeria has come full circle in its mitigation measures of human and climate challenges as it is now reinforcing cleavages in society. Conflicts exist within the goal and actions of government owing to narrowing of the playing field between 'hunger profiteers' and army of 'accumulation by dispossession' advocates. In the coming years, the battle field is no longer between fossil fuel apostles but with those who are bent to subvert every situation to their private end. Soon the nation (to borrow Paul Collier's phrase) is in for a double shock: oil windfall running out and restive populace faced with new mutations of the existential threat. When social and economic differences reinforce each other, it will make conflict more intense and society more polarised. In some cases, it has not just resulted to degradation but the end of way of life. Today, economic war will reinforce clearages since it is between forces within the country pulling at different ends but tomorrow's conflict will be as result of cross-cutting clearages when differences pull people in diverse directions, instead of reinforcing them. The latter will be with those who squandered the future generation's environmental parsimony on the altar of greed. In all these conflicts the environment is immediate loser while the people are the ultimate losers. Regardless, armed conflicts have never succeeded in upturning debauchery of nature. The major success is in breeding a new army of rent seekers to replace the old class. Statistics bear out in Nigeria and other African commodity frontiers like DR Congo, Angola and South Sudan that cultural transformation is not a day's job because mindset formed over many years are not changed overnight. But a day of reckoning comes and this reckoning came from an unexpected quarter: COVID-19. Nigeria, sorry to say, is not rooted in long-held concerns about planning for unforeseen.

Despite repeated efforts to correct this gap, the void is growing daily. Increasingly, the actual number of people living below the poverty line is growing. Statistics from National Bureau of Statistic in May 2020 points to 82.9 million Nigerians live on less than $1 a day representing 40% of the total population. Again, 52% of people in rural areas live in poverty compared to 18% in urban areas. A reason is failure to diversify the economy and build much-needed transport and power infrastructure which has stymied growth and ensured the spread of wealth only to rich elite.

The problem with the environment crisis is who will really speak for the people. The CSO sees the environment through ecological lens while the political elite use personal gain barometer to gauge success of policies. All these forces have one goal: sustaining group interest thus are amplifying considerations with group gain at heart. With the high visibility of political elites at national level their class preferences always hold sway. Not surprisingly, seemingly well intention policies have often backfired due to absence of political will. It has become increasingly clear not just to economists but the ordinary man that Nigeria is playing a giant game of economic whack-a-mole. In this endgame, the country will find a way of fixing one problem only for another potentially greater one to emerge. The nation it may seem is in one big giant merry-go-round roller coaster trip lacking any significant movement just like the barber's chair, it is all motion but no movement. Clearly, then, a developing country like Nigeria is like a small boat in a high sea subject to the vagaries of the elements.

The New Normal

The world is at a critical stage where issues of existential threats fallout of the coronavirus pandemic and economic meltdown are pushing the world community towards isolationists' policies at the same time with the need for multilateral interventions. This is another period when we must give hope to one another. In this new age, we either accept to face nature's degeneration together or confront nature's devastations without one another. We have the option either to rescue this and future generations from a life of greed and destruction of human and non-human nature. As this generation placed the planet in harm's way through our lifestyles, we have murdered sleep hence cannot go back pretending all is normal. What is normal today being an aberration yesterday? In this post normality society, there is a loser and winner. Humanity is the main losers while the major gainer, of course, is nature. These days what was normal in the past of wanton depletion and exhaustion nature is now our peril.

What is normal is not normal but figment of our imagination. Brene Brown sums this presciently when he states: We will not go back to normal. Normal never was. Our pre-corona existence was not normal other than we normalized greed, inequality, exhaustion, depletion, extraction, disconnection, confusion, rage, hoarding, hate and lack. Hence Brown enjoins us not to long for the past as it was never a perfect past. With the upsurge of coronavirus we are being given the opportunity to stitch a new garment—one that fits all of humanity and nature.

Indeed, we are living in a dystopian world ruled by democratic, totalitarian, capitalist and technocratic systems. In this new age, government control, environmental destruction, technological control, survival of the fittest and loss of individualism is high. The present condition is therefore a post-normal time which Ziauddin Sardar perceives an in-between period where old orthodoxies are dying, new ones are yet to be born and very few things seem to make sense.

The world in the post-COVID-19 Age is trying to make sense of the rapid changes taking place today. We have a new opportunity to stitch a new garment in ensuring we get the ecological disequilibrium right. Rightly, the equal exchange is an act of love for humanity not nature. But in loving nature we love the source of life by eradicating unequal relations either in socializations of nature or in our social relations of production. We have a new garment to stitch a path to human security. This garment leads to a path where eco-equilibrium relationship with nature is no longer a one-sided dialogue or monologue of the deaf initiated by environmental optimists. But one where there is coequal existence and harmonious relationship with nature.

Reaching an equitable solution on limiting carbon emission in the post-Hobbesian dystopia where everyone is at war with themselves proved to be another visual delusion. Unarguably, stalemate prone transition to the post-fossil future is driven by captains of industry with other motives for nature. Regrettably, these same captains are the main sponsors of major abatement policies at international conferences. Where these same actors, who captured previous model interests, have been dominant, their total control of people and nature will turn the poor into remnants of history. That way, they have achieved wholesale control of the global economy, which was impossible using the back door. From the front door we are witnessing reinvigoration of the fledging wheels of monopoly capital. But it is not only common to Nigeria as other African states are also entrapped in the vicious grip of fossil capital.

Nowadays, we are witnessing supplanting of the interests of African nations with those of large bio-corporations in a new eco-colonial world order—A world order bent on reintegrating African states who escaped from the ambit of the influence of the West back into the capital's orbit. Even as partners there are still vast inequalities in profit sharing with African elites. The ensuing unequal relationship arising from this unholy matrimony widened the gap between rich and poor nations creating amiable conditions for further returns on investment.

The Post-Carbon/COVID-19 Future We Want

The most dramatic changes in Nigeria will occur in 2020. COVID has fundamentally changed the way we view nature and do business. It is redefining relations within and between nations creating multilateral and unilateral moments at the same time. COVD-19 is drawing humanity towards unilateralism while our common destiny is pulling us towards multilateralism. For all its faults, the coronavirus has not only exposed world community vulnerability but also our susceptibility to change. Most importantly, it placed human security squarely where it belongs: At the foreground of economic discussions. Before now little consideration was given to the stress felt in the oil sector and need for diversification. Neither was the profligacy of previous regimes an issue of concern. COVID-19 is feared more by the superrich while the poor fear hunger than coronavirus. The major gain is medical services as the rich who were prone to escape on medical tourism to India, UK, Dubai and USA found that their box load of dollar was powerless to rescue them from the clutches of COVID-19. There is a positive aspect of the virus as we shall soon be witnessing improvements in health infrastructure and higher payments to health practitioners who are at the national frontline of battle to end the pandemic.

Unfortunately, the economy is the sacrificial victim of our policy deficiency. The surprise about the recent recession is not so much that it happened but it caught planners of national development with pants down. The planners went about making plans for recession neglectful other extraneous issues could also stymie growth. The void created by expectation and reality is so high in Nigeria. Due to a combination of factors these expectations were not realized. Unfortunately, 2020 is here with us now what was seemingly a long time off has come and gone. Nevertheless, the expectations and reality gap is widening.

Taking this into account, there is still the consensus Nigeria's fossil fuel dependent economic system will continue to change the ecological space in the years to come but a net zero carbon emission trajectory is not really feasible. A net zero emission economy is the ultimate goal since it is a game changer capable of fostering a post-carbon civilization. Even so, implementing a net zero emission transcends current energy security regime. COVID-19 has put a temporal stop to the coffin of fossil fuel nailed by the new wave of adoption of low carbon technologies. It temporally regenerated imminent death fossil fuels infrastructure globally could face. This owes to the fact that the world in their haste to bury fossil fuel failed to exorcise ghost of hydrocarbon from the global energy economy. Hence chronically non-functional fossil fuel infrastructure is still accorded more value than efficient technologies in renewables.

There are too many externalities for the low carbon revolution to really take place. For one thing, we are not developing new technologies as we ought to. The world is not innovative enough to change the world in radical ways. Saving mother Earth is a laudable goal but require new political architecture based on adaptation and sustainability tenets. The new social contract is basically an agreement to change the national production, consumption and cultivation pattern. It is a new value system with strong linkages to the culture of attentiveness born of a sense of ecological responsibility-a culture of participation coupled with a culture of obligation towards nature and future generations with intergenerational equity questions yet to be answered.

For clean energy transition to take place Nigeria should therefore use her vast biomass capacity, most especially, the vast bamboo wealth to surmount decarbonisation commitments within the revised NDC resubmission in and post- 2020.Other renewables such as hydropower, solar and wind will still play central roles in elimination of the carbon intensity of the economy. Capacity building to improve skill is important in this technological shift because it determines the level of innovations in a country. The country should therefore build her technical capacity to face these technological innovations in wind and solar. To this end, all facets of the economy must embrace the low carbon drive. Women are a constituent part of such development and cannot be left out. Any technology to be introduce must be gender sensitive as studies show that the adoption rate of current technologies to address the problem is low due in part to its gender blindness without adequately considering women and children's needs. The gender agenda in addressing climate change should also take a centre stage at all climate discourses if we are sincere in addressing climate change as its impacts are also gender differentiated.

Translating the UNFCCC Gender Action Plan (GAP) into national context is another way to go in engendering gender agenda on climate change, a low carbon pathway and sustainable development. The implementation of the global GAP should also reinforce the National Gender Action Plan of the Federal Ministry of Agriculture and Rural Development (FMARD) and other gender initiatives being promoted by government, the private sector and NGOs in the country. Integrating women who sustain the family farmer's livelihoods into any blueprint for the future should be overriding goal of Nigeria's post-carbon society. This roadmap should also include widespread cultivation of bamboos with country specific challenges at heart. Such blueprints should be in line with internationally agreed standard set in the SDGs. The SDGs targets should be the baseline for implementation of Nigeria's decarbonization initiative. The green landscape is opening in the country with these activities which presents new business opportunities on an alternative source of energy and new sources of livelihoods. Investment in bamboo is one such venture capable of creating opportunities for public/private partnership. Investment in solar and wind should go hand in hand with bamboo cultivation.

It is pertinent to underscore that the low carbon development emerging from this energy-mix is not an 'open sesame' to green transformation, eco-efficiency and eco-integrity. Neither is it an end of Nigeria's carbon emissions. Carbon emission will not end with mere deployment of new technologies or the fear of crude oil running out or that it spells economic doom neither will COVID-19 continually determine energy consumption. Rather, the end of carbon would be based on the realization that in addition to its beneficial attributes carbon emissions can also have severe consequences for humans and the environment -our life support. As such an awareness of adverse effect of our carbon profligate lifestyle and destruction of endanger species must be made. Deployment of renewable energy and fossil fuels production must therefore grow at the same rate in the near future until there is an eco-equilibrium between consumption and production of both hydrocarbon and biomass or when fossil fuel will inevitably run out or is phased out. This requires a conscious effort aimed at energy efficient alternatives.

To promote energy efficiency, the country needs both technical and financial assistance to chart a path best suited to its natural resources circumstances. The nation should be financially encouraged to adopt environmentally efficient technologies within the renewable sector in relation to its economic and ecological circumstances. Government should be supported to use the most affordable route within the country's human capability. But care must be taken cultivation of bamboos and uptake of new technology avoided the eco-costly neo-colonial era appropriation of nature where there is food crop and feedstock complex.

A new political architecture to drive change to low carbon growth path should explore new ecological capabilities to stimulate eco-development through eco-efficiency. Within this framework, humans and nature are harmoniously united in a steady-state. These eco-equilibria state should be devoid of marginalisation insisting on efficient use of nature's free gifts through a system of dynamic interactions and avoiding the tragedy of the Commons. It is not to deny problematic nature of eco-sufficiency or consequential marginalisation of some people. Rather, such development should be based on availability of environmental policy incentives essential for improvement of human welfare which also include health challenges. In a country enmeshed in a plethora of challenges, substitution among various forms of capital is not really feasible. A trade-off is inevitable given that maintaining the capital stock to meet human demands will promote single resource dependence. To avoid single resource dependence implies there should be some form of substitution among different forms of capital- a trade-off between biomass, solar and wind on one side pitched against fossil fuel. Therefore, the country should embrace co-fire or co-generation as a clear option for tackling issues of intermittency.

The key issue here is how to bring about synergy between diverse voices insisting on environmental salvation and those pushing for a pollution intensive pathway. With a weak market economy, fossil-led growth cannot be wished away neither can COVID-19 stifle its growth because it did not drop from a parachute in the sky. Rather, it is closely interrelated with other diverse issues. Not least of the factors is elite preference. Presumably, the wellbeing of the ordinary man is pivotal to the implementation of the NDC but in reality elite's preference is material to determining what is best suited for the poor. This explains why no mention is made in the ERGP of the instrument to reduce the carbon intensity of the economy in concrete terms. Instead, a vague reference was made to reduce gas flaring without quantifying actual amount of carbon it will reduce, where, when and how. The inability to specify in objective terms what should be reduced, how it can be implemented and the specific quantity to be achieved serves to reinforce the business-as-usual approach of gas flaring.

That said, present and future generations' welfare should not be politicized neither can it be bartered. While it is true there is a nexus between natural capital exhaustion and economic growth, the carbon debt of today should not jeopardize survival of humanity in the near future. Given there can be no future generation of Nigerians without present generation; we must aim for a policy with a level playing ground for socialization of nature. So, our utilization of nature must be with moderation in line with Africa's bio-communitarian world view wherein we are only keepers of our children's ecological wealth.

Despite these serious challenges asphyxiating the clean energy transition, hope is not totally loss. The journey to ecological neutrality is still bright as the future holds great promises for the energy revolution to take off. Our optimistic view does not mean policies for low carbon economy will automatically transform the country overnight. Not at all, rather, it shows how a promising sector can be optimized most especially biomass sector with bamboo cultivation if given higher priority to transform what is seemingly a negative into an economic positive. Not just because it has high environmental imperatives but mainly due to it favouring restoration of forest resources and can also regulate temperature in an area.

Against this backdrop, the Paris Agreement acclaimed a game changer in transforming global carbon budget is not the cure-all or harbinger of environmental salvation in some cases. The findings of IEA and IPCC in 2018 show even if the Paris commitments are met, it will not stave off serious anthropogenic climate change by 2050. So, the Paris commitments is not the last word on decarbonisation neither is it the only road to environmental salvation. In Nigerian context, Paris Agreement foreground cutting of carbon intensity of the economy with economic diversification envisaged by the country incapable of bailing the economy from looming danger. The problems ahead for the nation is whether the warring political and economic gladiators at the national stage will allow a truly people-centric development or follow the business-as-usual 'drill-baby-drill' archaic approach with its army of stranded assets. Diversification in the coming years will be the outcome of the collective will of the people not as a prescriptive magical spread when apply to the economy then viola! An automatic metamorphosis instantly takes place. This is not the case. In fact, it is dependent on the people who are real change makers and how they are adequately equipped to own policy on transformative change. What is most likely to occur is there will be readjustment of people's demands and consumption to tally with changes in the world. With the impending slump of crude oil prices, the rush would be on to exploit agriculture and other renewables as a bailout of economic shocks to be experienced. With reinforcement of fossil-led growth, emission intensity of the economy is likely to increase in future. Evidently, government only succeeded in substituting one form of pollution for another. To overcome ongoing reinvigoration of carbon growth, external cost should be built into all forms of fossil-based degradations. Non-existent external cost pricing is a source of market failure exacerbating extreme climate events. Government can through emission taxes reduce carbon pollution in cities via disincentives for automobiles emitting gases. Some industries emit dangerous gases from generating electricity in power plants through burning gases and oil.

In the short run, these industries are characterized by marginal and average cost but large-scale production will reduce unit production costs and will be more efficient eventually. An increase in price of electricity should result in lower consumption, in turn, would mean a reduction in the quantities of fuel used hence less environmental externalities and an improvement in environmental quality. With the level of poverty recorded in the country, a high energy price could potentially have a negative impact since it can accelerate climate changing trends. A more efficient price costing structure wherein output of polluting firms determine emissions hence price is needed. Efficient price costing will require providing adequate information regarding costs of environmental damages. The existence of natural boundaries and separation of the market is a constraint to allocating costs. Therefore, sub-national organizations (states and local governments) should be charge with the responsibility of gathering information on transaction cost and coordinating transaction related problems.

Overcoming this shortcoming means institutional failure resulting from rent seeking activities that has reduced government's effectiveness is totally stamped out. The key to a prosperous future is for principle of democratic governance be truly enshrined, where policy agenda and implementation is people-oriented; wherein nature is both the object and aim of development. And as all activities have environmental nexus, combating poverty through limiting the deleterious effects of climate change, health challenges, food scarcity, energy insecurity and above all ecological scarcity will mean creating poverty reduction schemes. As Lovelock once said, overcoming poverty is not a task of charity; it is an act of justice. We are totally in agreement with Lovelock's assertion. If this is the case, then, the present development agenda must address root causes of poverty threatening our food security, access to water, ecosystem stability, natural disasters and human health. An equitable distribution of resources is highly desirable if the country should solve the impending crisis.

Creation of a "fee-and-dividend" system advocated by Hansen (2012) for fossil-fuel companies should be charged. An easily implemented carbon fee imposed at the well head, mine shaft, or point of entry, with 100% of the revenue collected distributed monthly to the people on a per capita basis as dividends is our solution to the corrupt prone fossil fuel sector. An approach with the advantage of charging 'fee and dividend' from a climate-change standpoint will make fossil-fuel companies pay, while increasing the price of carbon to decrease its consumption. The proposed 'fee and dividend' regime will bring positive change in society while also ensuring the poor are adequately empowered to contribute to greening the economy.

President Muhammad Buhari's Administration started in the right direction towards a socially inclusive growth with remarkable achievements in the agricultural sector. But it is yet to be seen how his electioneering campaign translates into concrete reality in the energy sector. Clearly, government cannot bear this burden alone the private sector must also key into government policies if we are to make any significant progress. Public Private Partnership is essential if the ultimate goal of poverty reduction and improved environmental quality must be attained. For now, a limited number of private sector investors are participating in the deployment of clean energy technologies. The people have not fully participated because of income imbalance. Sorry to say, the level of in equality within the country cannot propel growth to a new height where the vast renewables can be explored. Government intervention in income redistribution is vital to empower the people who will, in turn, channel investments into clean innovations and deployment of new technologies to tap these vast resources. To jumpstart the process requires active private investments. Small and Medium Enterprises (SMEs) have an important role to play in the exit from fossil fuels. The Non-Governmental Organizations' role in this transformation journey cannot be over-emphasized given that some of them have demonstrated commitment in the campaigns and implementation of fossil-fuel exit and partnering with private sector and other development partners to deploy renewable energy solutions in the country. The involvement of research institutions in this quest to drive 'the change we want' and role of universities is equally crucial. There is need for urgent review of our educational curriculum at all levels to strongly integrate practical measures and awareness on addressing the current and anticipated development challenges such as energy security, coronavirus and climate change among other contemporary issues of our time.

A carefully planned plan to leave fossil fuels calls for efficiency, output and profitability to feed the rising population, while tapping into vast opportunities of decarbonisation with the agricultural sector synergistically integrated into the plan. More so, bamboos' commercial cultivation can foot government's bill of curbing hydrocarbon dependence. Transforming the agriculture value chain is in line with the ATA, NARF and APP. Bamboos when efficiently utilized will not only act as food supplement but will generate additional revenue for smallholder farmers. When fully exploited, the vast forest resources can drive growth in this sector via increased utilization of bamboo such that domestic use, additional export earnings and increased vegetative cover are its main benefits. The want of institutional infrastructure on bamboo can impede operationalization of using it to exit fossil fuels. Capacity building, education, distribution of seedling, enabling institutional backbone and private sector finance will provide necessary impetus to explore the full potential of bamboo in limiting the carbon footprint of the country. That is, if used for biomass energy in efficient cooking stoves like the Save 80 clean stoves deployed in some regions.

Furthermore, large swathes of land offer immense opportunities for solar, wind, hydropower and geothermal energy generation. Thus far, fossil energy requiring government subsidies is more attractive for the mere fact that it is more stable or readily available. At present, fossil fuel energy sources are cost competitive than renewables. Cost competitiveness of fossil fuels will be outdated because prices of renewables are dropping daily with innovative technologies deployed globally. To reduce the carbon footprint, government can promote mixed sources of energy to tap into the abundant energy potential.

All said the spectre of electric vehicles is here with us. In this low carbon regime, end of the oil age is inevitable. So, Nigeria's plan to exit from fossil fuel in the post-fossil era is unavoidable. In the years to come, FES implementation will be the defining challenge in the country though this could to be a source of economic boom. It could also be an avenue for economic doom if there is an amiable environment for sustaining industries and building efficient transport systems. All these depend on if government can take advantage of the vast opportunities it presents. Mechanism for removing waste via charting a pathway towards eco-efficient and eco-effective resource use must be promoted. A circular economy with limited waste is the ideal economic model. In the past, model of development was one in which goods where produced and discarded to create room for further acquisition of new materials. Modern circular economy de-emphasises discarding materials but is slant towards regeneration, recycling and reuse of materials. For instance, in the transport sector a circular economy would emphasize fuel efficiency through biofuel utilization, light rail, stricter standards for passenger and cargo transport, improvement in urban transit, and purchase of hybrid vehicles with high fuel economy. Enshrining a clean transport pathway in the national economic blueprint is one medium to transform towards a circular economy. Linkages with bamboo cultivation could be established in circular economy since in bamboos there are no waste as all the resource is used for other by-products.

To breathe a new life into the economy, renewable energy is key and future of energy sovereignty. The primary concern here is a change in orientation, changing our reliance on fossil fuel for renewable energy. Within this framework, there should be a new ethic of stewardship: the ethic of sustainable consumption of human resources based on actual needs and aspirations; where earth's resources are not stretched but are within its capacity to regenerate. This ethic of sustainable consumption is contained in a new vision for the future we suggested in the book 'Pathways to a Green Economy in the post-Paris era called the '**Food Sufficiency Economy**' (FSE).

The FSE concept fills gap in operationalization of biomass economy as a FES. The concept's appeal lies in placing a barrier on corporate malfeasance and profligate squandering of resources. In this theorization, prejudice common to many societies is resolved through nurturing a socially inclusive growth. The goal is to plot a path towards imbibing environmentally sound principles like the African ecobiocommunitarianists' world view. Within this blueprint, there is a harmonious relationship between human and nature since all species have co-equal value. It is in line circular economy but a home-grown solution based on Africa's cultural values. Here, Nigeria's ecological requirements shall be met within the country's biophysical limits while ensuring eco-development does not totally derail its economic growth. Because the economy will need a steady inflow of energy to sustain food production, sustainable use of resources is inevitable if the society must move away from the present reliance on others for food and energy requirements.

FSE is a convergence of ecologically sound principles borrowed from food sovereignty concept, circular and sufficiency economy. More importantly, this blueprint aligns with agroecological perspective to bring holistic change to the socialization of nature. It is the building block for transformation of the agricultural sector. It seeks to improve productivity in livestock and farmlands rather than mere physical increases in cattle headcount and lands under cultivation. Another positive attribute of the blueprint is it ensures every society is at liberty to define and determine their food system while being mindful of nature's wellbeing. When this principle is applied cultivation of bamboo, it can heal wounds of environment and eradicate profit maximizing machinations of interest groups. As the people are key to ongoing cultural and technological shift, transition should start with the biomass sector most especially using bamboos to drive change in the value system.

The main objective of this compass is to achieve green growth through ecologic principles of efficiency, productivity, mitigation, and adaptation. Within the framework, sustainable food production is bedrock of any socioeconomic transformation. In the light of current challenges posed by climate change, a combination of agroecology practice with conservation agriculture, Climate Smart Agriculture (CSA) and regenerative agriculture offers a huge potential and act as lubricants and catalysts for sustainable development, poverty reduction while ensuring food security and greening the economy. Change in the value system is essential because tradition and values are essential ingredients of society and must, therefore, form a vital part of the Food Sufficiency agenda. To change present consumption pattern requires a radical change in tradition and values people hold sacrosanct. People's tradition to a large extent determine their behaviour which also informed how they make choices and take risks. In the blueprint, agricultural transformation will give birth to a new culture of natural resource stewardship to gradually replace current unsustainable production and consumption pattern.

Appendices

Appendix 1

Climate actions taken by the government so far as Non-Annex I country to limit climate change (See Appendix 1) threats include:

(1) Long before Special Climate Change Unit (SCCU) was created, a coordinating body for climate change related issues had been established – the Inter Ministerial Committee on Climate Change (ICCC). It comprises representatives of (almost) all ministries and since recently also non-governmental representatives. The ICCC acts as a technical and advisory network of the federal government. It is supposed to support government as a multi-sectoral expert panel to make the right decisions onclimate change related matters. ICCC's tasks are concentrated on the elaboration and discussion of climate change related technical documents with the aim to implement them in practice, including the National Communication to UNFCCC (FMEnv, 2003).

(2) Complied with Kyoto Protocol by issuing the First National Communication (FNC) to the United Nations Framework Convention on Climate Change (UNFCCC) in November 2003.

(3) Created within the Federal Ministry of Environment, a Special Climate Change Unit now upgraded to the status of a department charged with implementing the Convention and Protocol related activities.

(4) Declared gas flaring as illegal, and has made a commitment to end the practice by the year 2008. There is also the Natural Gas Flare Out policy as contained in the National Energy Policy.

(5) In 2009, the Nigerian House of Representatives passed a law on the formation of a National Climate Change Agency. Though the bill has still not been accented to.

    This new agency shall serve as an executive sub-branch of FMEnv.

(6) End of July 2009, the Nigerian Senate passed another law on the formation of a new institution dealing with climate change, the National Commission on

A. S. Okoh, *Oil Mortality in Post-Fossil Fuel Era Nigeria*,
https://doi.org/10.1007/978-3-030-60785-2

Climate Change. Although the National Climate Change Commission Bill (see Appendix 1) and the Bill on Special Ecological Agency has been passed by the National Assembly it is still pending harmonization of the two chambers of legislature and the President's signature.

(7) The Senate Committee on Environment is currently engaged in formulating enabling law to create a commission whose role is to educate the public to make them aware of climate change and its effects on Nigeria.

(8) Created various agencies such as the National Institute of Freshwater Fisheries Research; National Centre for Arid Zone Studies; National Center for Climate Change; Nigerian Institute for Marine Research; and the Hydropower Development Department of the National Electric Power Authority.

(9) Enacted the following legislations: National Policy on Environment (1999); National Guidelines and Standards for Environmental Pollution Control in Nigeria (1991); National Effluent Limitation Regulation (1991); Waste Management Regulations; Pollution Abatement in Industries and Facilities Generating Waste Regulations of 1991; Environment Impact Assessment Decree No 86; Natural Resources Conservation Action Plan.

(10) Has given the responsibility of climate research and observation to the Nigerian Meteorological Agency.

(11) National Adaptation Strategy and Plan of Action for Climate Change in Nigeria (NASPA -CCN) has been designed to respond to an urgent need for a coordinated and integrated approach to systematically and effectively tackle the threat of climate change.

(12) Nigeria also submitted the Second National Communications (SNC) in February 2014.

(13) In October 2014, the FEC awarded a contract for N9.2 billion worth of cooking stoves and wonder bags for rural women under the National Clean Cooking Scheme. In the contract, it was agreed 750,000 units of clean stoves and 18,000 units of wonder bags were to be supplied by the designer from South Africa.

(14) Nigeria's active participation at regional and global meetings on climate change.

(15) Constitution of the Inter-ministerial Committee on climate change and the inauguration of a Roundtable Committee on climate change in August 2009. The Roundtable provides the opportunity to coordinate various activities of stakeholders through information sharing and collaboration to ensure greater efficiency and effectiveness.

(16) Other policies related to climate change mitigation include: National Policy on Drought and Desertification; Great Green wall for the Sahara and Sahel Initiative (GGSSI); National Biodiversity Strategy and Action Plan (NBSAP) Drought Preparedness Plan; National Policy on Erosion; Flood control and Coastal Zone Management, National Forest Policy; National Biodiversity Strategy an Action Plan.

(17) Establishment of National Conservation Strategy (NCS), Federal Environmental Protection Agency (FEPA), and the National Environmental Action

Plan (NEAP) andthe National Conservation Strategy and the National Environmental Action Plan. These were Nigeria's main strategies on sustainable development focusing on the conservation of natural resources and biodiversity.

(18) Constituted the Advisory Committee on Agricultural Resilience in Nigeria (ACARN); inaugurated the National Agricultural Resilience Framework (NARF) to advise government on the policies required to successfully implement a national climate smart agricultural programme (FGN, 2015). Government also crafted the Agricultural Transformational Agenda (ATA) in order to unlock the full potential of agriculture.

(19) Formulated NSRCC policy to mitigate climate change.

(20) Submitted Intended Nationally Determined Contributions to COP21 in October, 2015 at Paris.

(21) Ratified Paris Agreement in April, 2016 at New York.

Appendix 2

Projects, policies/initiatives on climate change mitigation in Nigeria

Initiative/Policy/Project	Objective	Value/source	Type of fund	Remarks
A local community processes waste sawdust from lumber mills into high-quality fuel briquettes in Lagos	To provide eco-friendly solution to the waste management of sawdust, offer alternative clean energy source and generate revenue for the local community	Not stated Global Environment Facility	Small Grants Programme	SEED award winner in 2011
The Affordable Energy Solutions Ltd which is a renewable energy company that provides solar panels, inverters, deep cycle batteries, charge controllers to local communities	To harnesses the power of solar energy to offer clean alternative energy to traditional fuels especially to rural and semi-urban communities	Not stated Self-funding	N/A	Affordable Energy Solutions Ltd has been recognised through awards for best renewable energy engineering company at the African Governance & Corporate Leadership Awards of Nigeria's Institute for Government Research & Leadership Technology

(continued)

(continued)

Initiative/Policy/Project	Objective	Value/source	Type of fund	Remarks
White Paper on Renewable Energy	Access to energy for rural communities far from the electricity grids; reduction in CO_2 emissions; diversification of electrical supply and energy security; promotion of the development of small, medium and micro enterprises (SMMEs)		PPPs	
Solar Nigeria Programme	Improve welfare outcomes of the currently underserved communities in Lagos State and Northern Nigeria by making a significant financial contribution towards the solar power electrification of public institutions	£66,062,369 IBRD (HSBC)		Project is at 44.54% completion by budget
Jigawa 80 MW photovoltaic solar power plan		$ 170 million Nova Scotia Power Development	CDIL, AfDB	
Partial risk guarantee program in support of the power sector				
Mainstream Hydro Project	For Mainstream to recover available capacity from a combined total of 650–1,338 MW	USD 134,586,000 ADB, Delta		African Development bank group
GreenWish to Invest $280 Million in Nigeria Solar Plants	To build solar power plants in Nigeria that are expected to start producing electricity	$280 million GreenWish		
The Rural Women Energy Security (RUWES) Project				
Scaling up Small Hydro Power (SHP) in Nigeria		19,939,680.00 USD GEF Trust Fund	Grant UNIDO	

(continued)

(continued)

Initiative/Policy/Project	Objective	Value/source	Type of fund	Remarks
Sustainable Fuelwood Management in Nigeria	Secures multiple environmental and socio-economic benefits, including reduced GHG emissions from wood fuel consumption, enhanced carbon storage and sequestration	20,930,000.00 USD GEF Trust Fund	Grant UNDP	Focal Areas in climate and land degradation
De-risking Renewable Energy NAMA for the Nigerian Power Sector	Support the development and implementation of a NAMA in the energy sector, namely a RE NAMA for the Nigerian Power Sector (NPS)	218,080,000.00 US GEF Trust Fund	Grants UNDP	
SPWA-CC: Mini-grids based on Renewable Energy (small-hydro and biomass) Sources to Augment Rural Electrification	To develop policy and a conducive market environment in order to promote renewable energy based mini-grids for augmenting rural electrification and productive uses in Nigeria	14,616,800.00 USD GEF Trust Fund	Grants UNIDO	
Rural Electrification and Renewable Energy Development	To support the design and implementation of several access expansion, intensification and renewable energy pilot projects	10,000,000.00 USD The World Bank	IDA	
Compressed Natural Gas (CNG) Vehicles		NIPCO		
Solar IPPs		$ 1.5 billion USAID, Nigerian Government		

(continued)

(continued)

Initiative/Policy/Project	Objective	Value/source	Type of fund	Remarks
Azura-Edo Thermal Power Station		US $860 million Amaya Capital Limited, Aldwych International Limited, African Infrastructure Investment Fund 2, Asset and Resource Management Company Limited, Nederlandse Financierings Maatschappij voor Ontwikkelingslanden N.V.("FMO"), and the Edo State government of Edo State, Nigeria		
NAIJA LIGHT Solar Electrification Programme		$ 1.2 million Federal Ministry of Environment Indian Government		
National Policy on the Environment	It presented specific guidelines for achieving sustainable development in fourteen sectors of Nigeria's economy	Government, private and donors	Excheque	There are range of policy instruments used by the government in Nigeria to promote sustainable production and sound environmental management. The instruments have been classified into market based instruments (MBI), regulatory instruments, negotiated agreements and information-based instruments

(continued)

(continued)

Initiative/Policy/Project	Objective	Value/source	Type of fund	Remarks
Africa SME Programme LOC—AB Microfinance Bank/Nigeria Limited	In order to support the expansion of ABN to the national level with a significant increase of its SME portfolio, ABN will require ap proximately USD 20 million to be sourced by term deposits, and medium and long term loans from financial institutions			
Inclusive Basic Service Delivery And Livelihood Empowerment Integrated Project		UAC 197,375,833 ADB, ADF, RWSSI, Government		
Indorama Fertilizer	To increase employment opportunities, enhance community development and reduce bottlenecks of supply chain distribution of fertilizer products	US $795,252,000 ADB, Cofinancier		
Nigeria Innovations Cluster Mapping Project	To nurture home-grown businesses and innovation effectively; To uncover the pockets of industry clusters where companies with similar attributes coexist in Nigeria	N/A The Tony Elumelu Foundation African Institute of Technology (AFRIT)	Grants	
NERICA	To improve food security and reduce poverty To increase national rice production	US $34 million Africa Rice Centre		
MIC GRANT STRENGTHENING OF FEDERAL MINISTRY OF AGRICULTURE AND RURAL DEVELOPMENT		USD 510,993 MICF, Delta	Grants	

(continued)

(continued)

Initiative/Policy/Project	Objective	Value/source	Type of fund	Remarks
Food-IAP: Integrated Landscape Management to Enhance Food Security and Ecosystem Resilience in Nigeria	To Contribute to Enhancing Long-term Environmental Sustainability and Resilience of Food Production Systems of Nigeria to Achieve Improved National Food Security	64,339,450 USD GEF Trust Fund	Grants UNDP	
GGW: Nigeria Erosion and Watershed Management Project (NEWMAP)	To reduce vulnerability to soil erosion in targeted sub-watersheds	508,592,593 USD Multi Trust Funds The World Bank		
LCB-NREE: Nigeria Child Project: Comprehensive and Integrated Management of Natural Resources in Borno State	To maintain the provision of ecosystem services in Borno state by preserving agro- and forest ecosystems in a context of improved production, conservation and renewable energy	35,838,929 USD GEF Trust Fund African Development Bank		
Plateau State Potato Value Chain Support Programme (PS-PVCP)		UAC 10,000,000 ADF		
Fund for Agricultural Finance in Nigeria (FAFIN)		USD 54,429,000 ADB, Delta		
Nigeria Agricultural Transformation Agenda Support Programme	To increase on a sustainable basis, the income of smallholder farmers and rural entrepreneurs that are engaged in the production, processing, storage and marketing of the selected commodity value chains	UAC 113,540,000 ADF, Nigerian Government		

(continued)

(continued)

Initiative/Policy/Project	Objective	Value/source	Type of fund	Remarks
Enable Youth Nigeria Programme	To create business opportunities and decent employment for young women and men along priority agricultural value chains of the various enterprises	US $200,600,400 ADB, Cofinancier		It is a national project
Lagos Urban Transport Project	Developing Pollution free transportation infrastructure		PPP	
Less Burnt for a Clean Earth: Minimization of Dioxin Emission from Open Burning Sources	Enhance human health and environmental quality by reducing releases and exposure to unintentional POPs originating from unsustainable waste operations	23,960,737.00 USD UNDP		
SPWA-CC: Promoting Energy Efficiency in Residential and Public Sector	To improve the energy efficiency of selected end-use equipment from the residential and public sector	9,829,532.00 USD UNDP National Energy Commission of Nigeria		
Unified Recycling Project	To reduce littering from PET Bottles; To create alternative employment and raise the income earnings of the citizens	N/A Coca-Cola		
GGW: Nigeria Erosion and Watershed Management Project (NEWMAP)	To reduce vulnerability to soil erosion in targeted sub-watersheds	US $508,592,593 Global Environment Facility (GEF), A Cofinancier		Focal Areas are in Biodiversity, Climate Change and Land Degradation
SPWA-CC: Nigeria Urban Transport	To improve the capacity to manage the transport sector in the Lagos and Kano metropolitan areas	329,500,000.00 USD The World Bank		
Environmentally Sound Management and Disposal of PCBs	Reducing Nigeria's Population exposure to PCB Effects and Associated Risks	41,746,612.00 USD GEF Trust Fund UNDP		Focal Areas are in Chemicals and waste

(continued)

(continued)

Initiative/Policy/Project	Objective	Value/source	Type of fund	Remarks
Sanitation, Hygiene & Water in Nigeria (SHAWN), Phase 2	To scale up water, sanitation and hygiene coverage in Nigeria	£96,519,997 UNICEF		It has attained 42.86% completion by budget
Nigeria Urban Transport Project: Abuja Facilitating Urban Mobility	To develop good quality public transport system in the city; To reduce reliance on cars and provide affordable transport	US $158.10 Million ADF, Clean Technology Fund (CTF)		
GEF National Portfolio Formulation Document	To enable the country to conduct a broad consultation process, to set priorities for GEF resource programming covering the 5th replenishment period	$30,000 GEF Trust Fund		
National Capacity Needs Self-Assessment (NCSA) for Environmental Management	To identify, confirm or review priority issues for action within the thematic areas of biodiversity, climate change, desertification and persistent organic pollutants	US $225,000 GEF Trust Fund The World Bank		

Source GIGGA, 2018
Adopted from Speech by Bill Gates, Co-Chair of Bill & Melinda Gates Foundation, at the National Economic Council Aso Rock Presidential Villa, Abuja, Thursday March 22, 2018

Appendix 3: Priorities of Agriculture Promotion Policy

The new policy regime, tagged the **Agriculture Promotion Policy** (APP) Policy is founded on the following guiding principles, a number of which are carryovers from the ATA reflecting the strong desire for policy stability. New elements added reflect the lessons from the ATA, as well as priorities emerging from the aspirations of the Buhari Administration:

1. Agriculture as a business—focusing the policy instruments on a government-enabled, private sector-led engagement as the main growth driver of the sector. This essential principle was established in the ATA and will remain a cardinal design principle of Nigeria's agriculture policies going forward.

2. Agriculture as key to long-term economic growth and security—focusing policy instruments to ensure that the commercialization of agriculture includes technologies, financial services, inputs supply chains, and market linkages that directly engage rural poor farmers because rural economic growth will play a critical role in the country's successful job creation, economic diversity, improved security and sustainable economic growth.

3. Food as a human right—focusing the policy instruments for agricultural development on the social responsibility of government with respect to food security, social security and equity in the Nigerian society; and compelling the government to recognize, protect and fulfill the irreducible minimum degree of freedom of the people from hunger and malnutrition.

4. Value chain approach—focusing the policy instruments for enterprise development across successive stages of the commodity value chains for the development of crop, livestock and fisheries sub-sectors, namely input supply, production, storage, processing/utilization, marketing and consumption. Building complex linkages between value chain stages will be an important part of the ecosystem that will drive sustained prosperity for all Nigerians.

5. Prioritizing crops—focusing policy on achieving improved domestic food security and boosting export earnings requires a measure of prioritization. Therefore, for domestic crops, the initial focus in 2016–2018 will be expanding the production of rice, wheat, maize, soya beans and tomatoes. For export crops, the initial focus will be on cocoa, cassava, oil palm, sesame and gum Arabic. In 2018 onwards, the export focus will add on bananas, avocado, mango, fish and cashew nuts. Investments in closing infrastructure gaps to accelerate productivity and investment in these crops will also be sequenced to reflect capital availability and management attention5. Prioritizing crops—focusing policy on achieving improved domestic food security and boosting export earnings requires a measure of prioritization. Therefore, for domestic crops, the initial focus in 2016–2018 will be expanding the production of rice, wheat, maize, soya beans and tomatoes. For export crops, the initial focus will be on cocoa, cassava, oil palm, sesame and gum Arabic. In 2018 onwards, the export focus will add on bananas, avocado, mango, fish and cashew nuts. Investments in closing infrastructure gaps to accelerate productivity and investment in these crops will also be sequenced to reflect capital availability and management attention.

6. Market orientation—focusing policy instruments on stimulating agricultural production on a sustainable basis, and stimulating supply and demand for agricultural produce by facilitating linkages between producers and off takers, while stabilizing prices or reducing price volatility for agricultural produce through market-led price stabilization mechanisms (commodity exchanges, negotiated off-take agreements, extended farm-gate price under value chains coordination mechanisms, agricultural insurance, etc.).

7. Factoring Climate change and Environmental sustainability—focusing policy instruments on the sustainability of the use of natural resources (land and soil, water and ecosystems) with the future generation in mind while increasing agricultural production, marketing and other human activities in the agricultural sector.

8. Participation and inclusiveness—focusing instruments on measures to maximize the full participation of stakeholders including farmer's associations, cooperatives and other groups, as well as NGOs, CBOs, CSOs, development partners and the private sector. This places a premium on the role of these organizations or groups as agents of economic change in general and agricultural economy in particular, thereby drawing benefits from their policy advocacy roles as partners to and watchdog of government.

9. Policy integrity—focusing policy instruments on measures for sanitizing the business environment for agriculture, in terms of accountability, transparency and due process of law, ensuring efficient allocation and use of public funding and fighting corruption on all programmes involving public resources. This also applies to compliance with international commitments, protocols and conventions that Nigeria is a signatory to.

10. Nutrition sensitive agriculture—focusing policy instruments on addressing the issues of stunting, wasting, underweight and other manifestations of hunger and malnutrition with particular reference to the vulnerable groups, which include children under 5, nursing mothers and persons with chronic illness and disabilities.

11. Agriculture's Linkages with Other Sectors—focusing policy instruments on the connected relationship between agriculture and other sectors at federal and state levels, particularly industry, environment, power, energy, works and water sectors.

About the Author: Okoh, Austine Sadiq (Ph.D.) is a Benue State University, Makurdi, Nigeria trained green political economist, climate change and energy policy expert. As a climate change and energy policy expert, he has three books and several academic papers to his name while contributing to major global indices. Dr. Okoh is a technical expert of a UN intergovernmental body- International Bamboo and Rattan Organization (INBAR) based in China where he contributed to the standardization of bamboo for renewable energy. He is a member of the Green Economics Institute, Trinity College, Oxford University, UK who are also the publishers of his bestselling book: *Pathways to Green Economics in post-Paris era Africa.*

About the Book: Nigeria's petroleum-led economy is grappling with plans hatched by parties to the Paris Agreement to wean the global economy from its adverse greenhouse gases upload. Different countries have laid out concrete timelines to curb the carbon intensity of their economies. A consequence of phasing out oil is the proliferation of renewable energy alternatives leading to dwindling income accruing the country.

Nigeria's oil mortality is no longer a far-off chance. Finding avenues of resolving oil's mortality simultaneously ameliorating adverse effects of COVID-19 pandemic is the main task of this book. The mask of this task is how to extricate from immiserating growth where scarcity rent generated from consumption of fossil fuel is the engine of economic recovery while simultaneously tackling competing needs for economic diversification in line with the decarbonization/green economic recovery of the post-COVID-19 age.

References

1. Adebisi, A. (2014). Action plan for the large-scale deployment of renewable. Energy sources in Nigeria. A one-day sensitization workshop and public presentation of the2nd edition of renewable energy master plan organised by federal ministry of power at Rockview Hotel, Abuja.
2. Anyadike, R. C. (2009). Climate change and sustainable development in Nigeria: Conceptual and empirical issues. Paper presented at the enugu forum policy paper 10, enugu: African institute for applied economics.
3. Ayodele, S. (1985). The conflict of the growth of the Nigerian petroleum industry and 114 environmental quality. *Socioeconomic Planning Science,19*(5).
4. Bala, E. J. (2013a). Energy and sustainable development in Nigeria: Issues and prospects. Presentation made to Course 22, National Defence College of Nigeria, Abuja. Retrieved November 12, 2013.
5. Bala, E. J. (2013b). Status of energy access in Nigeria. Paper presented at the workshop on energy access and village inventory organised by the world energy council, Johannesburg. Retrieved February, 18, 2013.
6. BNRCC, (2011). *Building Nigeria's Response to Climate (2011).* Toward a national adaptation strategy and plan of action (NASPA) a consultative document for stakeholders review and discussion NEST.
7. Bhagwati, J. (1993). The case for free trade. *Scientific American,269*(5), 42–49.
8. Bhagwati, J. (2002). Coping with anti globalisation: A trilogy of discontents. *Foreign Affairs,81*(1), 2–7.
9. Bhagwati, J. (2004). *In defence of globalisation.* Oxford: Oxford University Press.
10. Cervigni, R., Rogers, J. A., & Henrion, M. (Eds.). (2013). *Low-carbon development opportunities for Nigeria.* Washington, DC: The World Bank.
11. Collier, P. (1988). Oil shocks and food security in Nigeria. *International Labour Review,127*(6), 761–782.
12. Collier, P. (2008) Oil, growth and governance in Nigeria. In P. Collier, C. C. Soludo, C. C. & C. Pattillo (Eds.), *Economic Policy Options for a Prosperous Nigeria.* New York: Palgrave Macmillan.
13. Cotula, L., Vermeulen, S., Leonard, R., & Keeley, J. (2009). 'Land grab or development opportunity?' agricultural investment and international land deals in Africa. London/Rome: IIED/FAO/IFAD.
14. Desonie, D. (2008). *Humans and the natural environment: The future of our planet.* New York: Chelsea House Publishers.
15. DFID-Department for International Development. (2009). Impact of climate change on Nigeria's economy. Final report.

A. S. Okoh, *Oil Mortality in Post-Fossil Fuel Era Nigeria*,
https://doi.org/10.1007/978-3-030-60785-2

16. Eleri, E., & Yahaya, A. (2014). Green deal Nigeria? Stumbling along the road from fossil fuels to clean energy. In Heinrich Böll Stiftung (Ed.), Greening the continent: Reflections on low carbon pathways. South Africa: Heinrich Böll Stiftung.
17. Emefiele, G. (2020). Turning the COVID-19 tragedy into an opportunity for a New Nigeria
18. Etuonovbe, A. K. (2011). The devastating effect of flooding in Nigeria FIG working week 2011 bridging the gap between cultures Marrakech. Morocco. Retrieved May, 18–22, 2011.
19. EPA-Environmental Protection Agency. (2015). Clean energy. United States environmental protection agency. Retrieved from https://www.epa.gov/cleanenergy.
20. FAO-Food and Agriculture Organization. (2016). Climate change and food security: Risks and responses. Retrieved from www.fao.org/publications.
21. FGN- Federal Government of Nigeria. (2011). *National adaptation strategy and plan of action*. Abuja: Federal Ministry of Environment, Climate Change Department.
22. FGN-Federal Government of Nigeria. (2012). Climate change and sustainable development in Nigeria. National Report for COP18/cmp 8.
23. FGN-Federal Government of Nigeria. (2015). *Approved Nigeria's INDC Submision.*Retrieved from www.unfccc.int/submisions/INDC/submissionpages/submissions.aspx.
24. FGN-Federal Government of Nigeria. (2017). Economic recovery and growth plan 2017–2020. Ministry of budget and planning.
25. Global Report on Food crises 2020. https://www.fsinplatform.org/sites/default/files/resources/files/GRFC_2020_KM_200420.pdf.
26. Hansen, J. (2012a). Game over for the climate, *New York Times*. Retrieved May 9, 2012 from https://nytimes.com 12/12/2018.
27. Hansen, J. (2012b). Storms of My Grandchildren's Opa. Retrieved December 17, 2017, December 13, 2012 from https://columbia.edu.
28. Heinrich Boll Foundation and Friends of the Earth International. (2015) Coal atlas: Facts and figures on a fossil fuel. Retrieved October 12, 2017, from www.Nigeriacoal.org.
29. IEA (International Energy Agency). (2009). 2009 energy balance for Africa. Retrieved from www.iea.org/stats/balancetable.asp?COUNTRY_CODE=11.
30. IEA (International Energy Agency). (2015). World energy outlook energy: Climate and change, a publication of international energy agency paris, France. Retrieved fromwww.iea.org
31. IEA (International Energy Agency). (2016). *World energy outlook 2016*. Paris
32. IPCC-Intergovernmental Panel on Climate Change. (2014). *Fifth assessment synthesi(longer) report*. Retrieved from IPCC: www.ipcc.ch/pdf/assessment report/ar5/.../SYR _AR5_LONGER REPORT.pdf.
33. IPCC-Intergovernmental Panel on Climate Change. (2007). Climate change 2007: Impacts, adaptation and vulnerability. Working Group II contribution to the fourth assessment report. Cambridge: Cambridge University Press.
34. Kale, Y. (2016)*.Unlocking Nigeria's Potential…growing through diversification.* A Paper delivered by the Statistician General of the Federal and CEO of National Bureau of Statistics at the 7th Stanbic IBTC Bank Investors Conference, Eko Hotels and Suites Victoria Island, Lagos. Retrieved February 23, 2016.
35. Karl, T. (1999). The perils of the petro-state: Reflections on the paradox of plenty. *Journal of International Affairs,53*(1).
36. Karl, T. (1997). *The paradox of plenty—Oil booms and petro-states*. Berkeley: University of California Press.
37. Khor, M. (2010). Challenges of the green economy concept and policies in the context of sustainable development, poverty and equity. In *The Transition to a Green Economy: Benefits, Challenges and Risks from a Sustainable Development Perspective. Report by a Panel of Experts to Second Preparatory Committee Meeting for United Nations Conference on Sustainable Development,* pp. 69–94.
38. Kuehl, Y., Henley, G., & Yiping, L. (2011). The climate change challenge and bamboo: Mitigation and adaptation, INBAR working paper 65, a publication of INBAR.
39. Kuehl, Y., & Yiping, L. (2012). Carbon offsetting with bamboo, INBAR working paper 71.

40. Lingo. (2015). The Paris Mirage—Reducing emissions while increasing them.www.leave-it-in-the-ground.org.
41. Madzwamuse, M. (2010). Climate Governance in Africa—Adaptation strategies and institutions cape town: Heinrich Böll Foundation Southern Africa. Retrieved from https://za.boell.org/web/publications-631.html.
42. Mathew, A. O., & Adeboye, B. F. (2016). Agricultural sector and economic development: The Nigerian Experience.
43. NBS, (2012). National bureau of statistics: Agriculture. Retrieved: May 15, 2018 from https://www.nigerianstat.gov.ng/sectorstat/sectors/agriculture.
44. NNPC-Nigerian National Petroleum Corporation. (2017). Nigeria to earn US $150m from Bio-fuel Initiative Annually. Retrieved December 17, 2017 from https://www.nnpcgroup.com/PublicRelations/NNPCinthenews/tabid/92/articleType/ArticleView/articleId/204/Nigeria-To-Earn-US-150m-From-Bio-fuel-Initiative-Annually.aspx.
45. Okoh, A. I. S. (2014). Post-kyoto protocol negotiations: Sunrise or sunset for economic growth in Africa? *Journal of Good Governance and Sustainable Development in Africa (JGGSDA),2*(3). Retrieved from https://www.rcmss.com.
46. Okoh, A., & I. S. . (2015). Sustainable development and adaptation strategies of the poor: A study of 2012 lower benue floods in Nigeria. *International Journal of Scholarly Research,1*(1), 1–15.
47. Okoh, A. I. S. (2015). Can Clean Energy Transition (CET) save Africa from Energy Poverty? Paper presented at Nigeria alternative energy expo held at shehu isa yar-adua international conference centre, Abuja, Nigeria. Retrieved October 14–16, 2015 from www.nigeriaalternativeenergyexpo.org/index.
48. Okoh, A. I. S (2017). 'Pathways to a Green Economy in post-Paris Africa. https://www.greeneconomicsinstitutetrust.org/books/pathways-green-economy-post-paris-africa/
49. Okoh. A. I. S. (2017). Makurdi flood: Disaster waiting to happen?https://www.linkedin.com/pulse/makurdi-flood-disaster-waiting-happen-sadiq-austine-igomu-okoh/.
50. Okoh. A. I. S, Mailumo, D. & Orokpo O. F. (2017). Nigeria's nationally determined contribution (NDC) in the transition to a low carbon economy. *Dutse journal of Economics and Development Studies (DUJEDS),3*(1), pp. 27–41.
51. Okoh, A. I. S. (2018). Editorial of carbon free Africa daily. https://paper.li/f-1377284124.
52. Oladipo, E. (2010). The global framework for mitigating climate change: A conceptual analysis. In O. C. Eze, & O. Oche (Eds.), *Climate Change and Human Security inNigeria*. Lagos: Nigerian Institute of International Affairs.
53. Paehler, K. (2007). *Nigeria in the Dilemma of Climate Change*. Retrieved February 12, 2016, from https://www.kas.de/proj/home/pub/33/2/dokument_id-11468/index.html.
54. RMRDC. (2003). Report of the multidisciplinary task force on wood and wood products sector. Raw Materials Research and Development Council Publication
55. Roche, M. Y., Ude, N., & Ofoegbu, I. (2017). True cost of electricity: Comparison of costs of electricity generation in Nigeria. A Report of Heinrich Boll Stiftung/The Nigerian Economic Summit Group.
56. RMRDC (2010). Report of the multidisciplinary task force on wood and wood products sector. Raw Materials Research and Development Council Publication.
57. Sachs, J. D., & Warner, A. M. (1997). Natural resource abundance and economic growth. *Harvard Institute for International Development*.
58. Sachs, J., & Tubiana, L. (2014). *Pathways to deep decarbonization: 2014 Report*. A Publication of SDSN & IDDRI
59. Sambo, A. S. (2009). *Strategic developments in renewable energy in Nigeria*. Third Quarter: International Association for Energy Economics.
60. Sanusi, L. S. (2010). *Growth prospect for Nigerian economy*. Retrieved April 20, 2018, from https://www.cenbank.org. (CBNgovernor).
61. SNC (Second National Communication). . (2014). *Nigeria's second national communication under the united nations framework convention on climate change*. Abuja: Federal Ministry of Environment.

62. Smil, V. (2006). In OECD global science forum: Conference on scientific challenges for energy research, Paris.
63. Smil, V. A. (2000). *Review of Energy Environment,25*(1), 21–51.
64. UNFCCC-United Nations Framework Convention on Climate Change. (1997). *The Kyoto Protocol*. UNFCCC–United Nations Framework Convention on Climate Change (2009), Copenhagen Accord. Decision 2/CP.15. Bonn: UNFCCC Secretariat.
65. Tarheripour, F., Hertel, T. W., & Tyner, W. E. (2009). *Implications of biofuels boom for the global livestock industry: A computable general equilibrium analysis*. Department of Agriculture Economics: Purdue University.
66. UNEP. (2011).*Towards a green economy; pathways to sustainable development and poverty eradication UNEP.* Retrieved from https://www.unep.org/.greeneconomy/Portals/88/documents/ger/ger_final_dec_2011/Green%20EconomyReport_Final_Dec2011.pdf.
67. Wikipedia. (2018). *Economy of Nigeria*. Retrieved May 7, 2018 https://en.wikipedia.org/wiki/economy.
68. World Bank. (2011). From growth to green growth: A framework. Retrieved from https://www.wds.worldbank.org/servlet/WDSContentServer/WDSP/IB/2011/12/07/000158349_20111207171314/Rendered/PDF/WPS5872.pdf